Wireless Networks

Wireless Networks

From the Physical Layer
to Communication, Computing,
Sensing, and Control

Edited by

Giorgio Franceschetti

Professor of Electromagnetic Theory,
University Fredrico II of Napoli, Italy

and

Sabatino Stornelli

CEO of Seicos and Selex-SeMa,
Finmeccanica Companies

AMSTERDAM • BOSTON • HEIDELBERG • LONDON
NEW YORK • OXFORD • PARIS • SAN DIEGO
SAN FRANCISCO • SINGAPORE • SYDNEY • TOKYO
Academic Press is an imprint of Elsevier

Academic Press is an imprint of Elsevier
30 Corporate Drive, Suite 400, Burlington, MA 01803, USA
525 B Street, Suite 1900, San Diego, California 92101-4495, USA
84 Theobald's Road, London WC1X 8RR, UK

This book is printed on acid-free paper. ∞

Library of Congress Cataloging-in-Publication Data
Wireless networks : from the physical layer to communication, computing,
sensing, and control / edited by Giorgio Franceschetti and Sabatino Stornelli.
    p.   cm.
  Lectures presented in Capri,Italy, fall of 2004.
  Includes bibliographical references and index.
  ISBN-13: 978-0-12-369426-3 (pbk. : alk. paper)
  ISBN-10: 0-12-369426-4 (pbk. : alk. paper) 1. Wireless LANs—Congresses.
2. Personal communication service systems—congresses. I. Franceschetti,
Giorgio. II. Stornelli, Sabatino.
  TK5105.78.W578 2006
  004.6'8—dc22

                                                            2005036150

British Library Cataloguing-in-Publication Data
A catalogue record for this book is available from the British Library.

ISBN 13: 978-0-12-369426-3
ISBN 10: 0-12-369426-4

For information on all Academic Press publications
visit our Web site at www.books.elsevier.com

Printed and bound by CPI Group (UK) Ltd, Croydon, CR0 4YY

Transferred to digital print 2012

A cocktail of control, computing
Together with software and sensing,
Considered the latest boom,
Was tasted by a man in a room
While wireless nets he was using.

The man was sipping the beverage
By using the wisdom of his age.
He said that something was missing:
A book taking care of the fixing!
And this was well worth his wage!

– GF

Contents

7

Wireless Networks and the Expected Next Revolution in Information Technology 289

Foreword

As president and CEO of the Italian industry holding FINMECCANICA, I am elated to introduce this book on wireless communication, which collects the edited lectures of a summer school sponsored by FINMEC-CANICA in the beautiful frame of the Capri island, Italy, in the fall of 2004. The book chapters are authored by prestigious names in the realm of the pertinent scientific arena.

As testified by its title, the book is characterized by an integral vision of wireless communication, from the physical layer to communication, computing, sensing, and control. I believe that this is the modern, right approach to present this material for the benefit of all the readers, spanning the academic world to the industrial environment. Information transfer is a key element in the structure of our society. The next step is its full integration with remote sensing, data access, real-time processing, and up to the final stage of far-away physical actions implementation.

The modern science was born with the experiments performed and the mathematical models developed by Galileo and Newton. The physical science emerged as a very important component of our society. Physicists were developing theories and models of natural phenomena, and a new segment of the civil society, the engineers, made use of above results and accomplishments to construct machineries, build up products, and provide services, thus usually rendering people's life more productive, rich, and enjoyable. But in the middle of the last century a novel and unique happening took place: engineers and not physicists created a novel branch of scientific knowledge, the communications discipline, and subsequently exploited its practical implementation. The first accomplishments were made by Shannon in the United States and Kotelnikov in Russia, creating the Information Theory and developing its implications and measurements concepts. Then, a full array of hardware components and software codes were developed and installed to synthesize wired and wireless communication networks, vessels of the information blood of our society. As an electronic engineer myself and president and CEO of FINMECCANICA, with its over 8,000 engineers and 10,000 technicians, we are all very proud of all this.

We are now at the step of a further quantum leap: the integration of communication with computer processing, sensors data, and far-away actions. This is the academic and industrial answer to the new impelling demands raised by conventional traffic jams, homeland security, and life quality: it is the Global Village dream made true.

The book I am presenting is organized in this spirit, moving a small but significant step along this modern vision of wireless communication.

Dr. Ing. Piero Guarguaglini
FINMECCANICA President and CEO

Preface

This book collects the lectures held on the beautiful island of Capri during the fall of 2004. A great advantage of books resulting from short schools or specialized symposia is that they present the latest information on the subject, usually in a very readable form. At the same time, these books may lack an exhaustive full treatment of the matter, and some duplication of the material presented in different chapters may be expected. It is the responsibility of the editors to minimize these shortcomings without possibly impairing the freshness of the presentation by changing a collection of lectures into a formal textbook. This is what we tried to accomplish with a careful revision of all chapters provided by the school lecturers, but without enforcing our style, viewpoint, and perspective upon their writings.

The book consists of seven chapters that, except the first one, are derived from the school lectures. Each one of these chapters coincides with the revised (by each author) version of the school notes with a final (soft) touch by the editors. The chapters derived from the lecturers are presented first.

Chapter 2 outlines what makes radio communications distinct from wired communications at the physical layer and highlights current trends in radio design. An overview of radio wave propagation and its impact on communication system design is given. The issues of fading and path loss and their impact on the range and reliability of communications are presented. Modulation and coding techniques commonly used in wireless communications are considered. The importance of diversity in achieving reliable data communications is stressed, and an overview of techniques used to achieve diversity is provided. Radio architectures used in modern communication systems are discussed.

Chapter 3 is devoted to the receiving element of the wireless channel: the handheld, wearable, and implantable antennas used in personal communication technology. This is a very important issue, because their design should account for the electromagnetic interaction between the antenna and the human body, a key factor to be considered. A full array of results on this subject is presented: popular antenna designs (such as monopole, inverted F, reconfigurable patches, etc.), numerical techniques for antenna performance evaluation, electromagnetic exposure of people to handheld receiving devices, Specific Absorption Rate (SAR) for adults and children, etc. The chapter is characterized by a large amount of first-hand experimental results on innovative realizations and related tests performed at the UCLA Electrical Engineering Department.

Chapter 4 consists of an overview of the models of the wireless channel, including both numerical procedures and analytical results. The former can be used for the analysis of a specific built-up scenario and can lead to the generation of electromagnetic solvers, where ray tracing techniques, including reflected, diffracted, and creeping rays, are implemented. Solvers complexity is discussed, highlighting their advantages and limitations; details and results related to a particular solver are presented also. On the analytical side, the latest statistical techniques are introduced, ranging from random walk theory to innovative percolation models of the urban scenario. Applications to the evaluation of expected values of electromagnetic quantities of interest (as, for instance, the path loss) are presented, with analytical, numerical, and experimental results.

Chapter 5 presents an overview of ad hoc wireless networks, the kind of wireless technology that enables untethered, wireless networking in environments where there is no established wired or cellular infrastructure (e.g., battlefield, disaster recovery, homeland defence) or where it is not cost effective to use an existing infrastructure (e.g., personal networking, collaborative computing). All their characteristics, such as mobility, multihopping, self-organization, energy conservation, scalability, security, etc., are introduced. The challenges of the network layer — routing and multicast — are discussed, providing all the necessary details with reference to the areas of sensor networks, automated battlefield, and collaborative computing. The MINUTE-MAN (Multimedia Intelligent Network of UnatTEnded Mobile AgeNts) project, developed at UCLA, is presented as a case study, including simulation experiments and their discussion.

Chapter 6 deals with acquisition, processing, compression, communication, and reconstruction of real-world signals, like sound and video, in a distributed environment. A unified treatment of data representation, routing, and node placements in a sensor network is presented, which sets in the right perspective the optimization of various metrics of interest, particularly energy efficiency and accuracy of data reconstruction. Challenging questions, including fundamental, algorithmic, and practical issues, are addressed, with reference to lossless and lossy coding, optimal node placement problem, data gathering, and total distortion assessment. A relevant feature of this chapter is the treatment of the tight connection between data structure and transport mechanism, considered as the central challenge in the design of operational sensor networks.

Chapter 7 describes the latest results in wireless networks, providing an overview on the possible future scenario. Scaling laws are discussed, both with respect to current technological efforts and to the ultimate information-theoretical capability. Optimal architectures for information transport, protocol design, power control, medium access control, and routing are addressed. Nodes that can sense, compute, and wirelessly communicate are examined in detail. For such networks the problems of communication, sensing, and data fusion are inseparable; this next phase of the information technology revolution — the convergence of control with communication and computing — is examined. Even the implementation of far-away actions is briefly touched. The architectures of these widely integrated network systems and the prerequisites for their proliferation are addressed.

Examination of the content of the book shows that all the trajectories of wireless communication (as anticipated in the book title) are included. Coding and protocols; electromagnetic channel models; communication and sensors networks, as well as their integration; critical design issues to minimize the exposure to the electromagnetic fields while using the receiving devices; recent applications to homeland security and disaster recovery and mitigation; and the present and the possible future of this research area are all somewhere available in these chapters, linked by a logical progression. This unified treatment is unique, introducing the reader to this intriguing and fascinating world and providing a valuable scenario with all of its facets. Accordingly, the scientific and technical issues are all well covered. It seemed appropriate to also add an introductory chapter (whose material was not presented during the Capri school) outlining the impact of these emerging technologies on our life, and their shaping and addressing by our society needs. This short introduction, presented in Chapter 1, clearly does not pretend to be complete and rigorous. It only tries to set all the material of the subsequent chapters in the frame of its applications context, presenting the players of this complicated scenario and anticipating the possible future developments.

Some few considerations about the potential audience of this book may be appropriate. It is easily anticipated that the book may be useful to the large variety of scientists and technicians working in the area of wireless communication. Each category will find updated information about its specific application area, with the additional advantage of being also exposed to its connections with the nearby areas. This was the spirit of the school, and it has been saved in the book, also.

It might be concluded that the book audience is limited to professionals in the specific covered field, but this is not completely true. As discussed at the end of Chapter 1, the new, cross layer, interdisciplinary design paradigm in the communication networks area requires not only in-depth expertise in one specific layer, but also a well rounded scientific background that integrates, among others, all the disciplines touched in the book. It is responsibility of the academia to realize a program that can educate and form the *telecommunications engineers* of the future. It can be anticipated that an array of new *wide band* courses should be designed and implemented. The book we are presenting is certainly not a textbook by itself, but it might be initially integrated by notes for such a task, leading eventually to a more formal textbook realization. We conclude that an additional non-negligible audience is expected in the universities and among professors and graduate students.

The presentation of a book is usually concluded with appropriate acknowledgments to all those who helped its realization. But our acknowledgments are toward colleagues that provided much more than significant help, because they essentially wrote the book. Accordingly, we are deeply indebted to Professors Michael Fitz, Mario Gerla, Yahja Rahmat-Samii of UCLA, P. R. Kumar of UIUC, Massimo Franceschetti of UCSD, Martin Vetterli of EPFL and UC Berkeley, and Daniele Riccio of University Federico II Napoli: they transformed their lectures into chapters, some of them with the help of their collaborators Cong Shen, Michael Samuel, Zhan Li of UCLA, and Razvan Cristescu of Caltech. Our appreciation is also for the Elsevier staff, in particular, for the Assistant Editor, Rachel Roumeliotis, and the Project Manager, Brandy Lilly, who continuously encouraged and pushed us to complete our job. We also thank FINMECCANICA, who sponsored the school, and all the participants, for their comments and suggestions.

We can certainly state that this book has been a cooperative job. It is very likely that everyone should thank each other. Our personal view is not only that cooperation is the right way to proceed, but it should even be extended, especially in these research areas that exhibit such an important connection with the organization of our society. As noted in Chapter 1, there is increasing attention of national and international authorities to assure a friendly and safe environment to the social community, which implies early time knowledge of emergence situations, follow up in real-time of the evolving scenario, and final evaluation and intervention by using a decision making system, usually remotely located. These requirements may be fulfilled by the appropriate use of information and communication technologies, along the integrated view presented in this book. It is very desirable that a permanent exchange of information, about technological offerings and expected requirements, occur between the scientific community and all those public authorities whose officers are aimed at assuring a safer and most secure environment to all of us. We will say with Keats that this is

> *A hope*
> *beyond the shadow*
> *of a dream.*

[From "Endymion" by John Keats]

And to all these officers this book is dedicated.

The Editors

About the Authors

The short biographies are organised according to the Authors appearing from Chapter 1 to 7

Chapter 1

Giorgio Franceschetti was born and educated in Italy. Winner of a nationwide competition, he was appointed professor of Electromagnetic Theory at the University Federico II of Napoli, Italy in 1969, position that he holds since then. He has been Fulbright Scholar and Research Associate at Caltech, Visiting Professor at the University of Illinois, at UCLA, at the Somali University (Somalia) and at the University of Santiago de Compostela (Spain). In addition to his Chair at the University Federico II in Italy, he is currently Adjunct Professor at UCLA, Distinguished Visiting Scientist at JPL and Lecturer at the Top-Tech Master of University of Delft, The Netherlands, n Satellite Navigation. Author of over 150 (refereed) papers on Journals of recognized standard and nine books, and recipient of several awards, he is active in research on Electromagnetic Theory and Applications, Signal Processing, Synthetic Aperture Radar Imaging and Electromagnetic Propagation in complex media. He is Life Fellow of IEEE and Member of the Electromagnetic Academy. He recently received the gold medal from the President of the Italian Republic for his achievements in culture and science.

Sabatino Stornelli was born in Avezzano, Italy, in 1957. He graduated in Electrical Engineering at the University of L'Aquila, Italy, in 1982, with subsequent post-graduate additional studies at the Universities La Sapienza of Roma (Italy) and of Portland, Oregon (USA). After some industrial experience in Italy, he joined the European Space Agency (ESA) from 1987 to 1991, with R&D responsibilities in the Department of Information Technology. Coming back to Italy, he resumed his industrial activity, from 1991 to 1995 in Dataspazio and from 1995 to 2003 in Telespazio, operating in the area of Space Systems and Earth

Observation: his responsibilities steadily increased, reaching the level of CTO in 2000 and then General Director in 2001. Finally, in the year 2004 he was appointed CEO of Seicos, and in 2006 of Selex-SeMa, both of them Finmeccanica Companies operating in the area of Telecommunication Services, positions that he holds since then. In addition to this intense industrial activity, he has been Contract Professor at the Department of Aerospace Engineering of the University La Sapienza of Roma, Italy, in the years 1999–2000; he presented over 20 papers at International Symposia in the area of Satellite Navigation, Spacecraft Control and use of satellites for Environment Risks Management. In 1996 he got an Award from the Ukraine Academy for his activity in the area of Definition, Design and Management of Space Missions.

Chapter 2 [Course Lecturer: Prof. Michael Fitz]

Michael Fitz was born in Akron, Ohio (USA) in 1960. He obtained his Bachelor of Engineering Degree in Electrical Engineering from University of Dayton in 1983, and his M.S. and Ph.D. degrees in Communication Science from the University of Southern California, in 1984 and 1989, respectively. Dr. Fitz has been a professor at Purdue University, the Ohio State University, and the University of California Los Angeles (UCLA), and has held numerous positions in the private sector where he is currently employed at Northrop Grumman in Redondo Beach, CA. He was a recipient of the IEEE Communications Society Leonard G. Abraham Prize Paper Award. He is a co-author of the soon to be published book "A First Course in Communication Theory".

Cong Shen was born in China on July 1980. He obtained his B.E. and M.S. degrees, in 2002 and 2004 respectively, from the Department of Electronic Engineering, Tsinghua University, China. After that he joined Wireless and Networking Group at Microsoft Research Asia (MSRA) as a full-time visiting student in 2004. Currently he is pursuing his Ph.D. at Electrical Engineering Department, UCLA. His research interests include information theory, wireless communications and networking. He is IEEE Student Member.

Michael Samuel was born in Cairo, Egypt in 1976. He obtained his B.Sc. degree in Electrical Engineering from Ain Shams University, Cairo, in 1999 and his M.Sc. degree in Communication Technology from the University of Ulm, in 2003. He is currently pursuing his Ph.D. degree at the Electrical Engineering Department, UCLA.

Chapter 3 [Course Lecturer: Prof. Yahya Rahmat-Samii]

Yahya Rahmat-Samii was born in Tehran, Iran. He obtained his B.S. Degree in Electrical Engineering from Tehran University and his M.S. and Ph.D. degrees from the University of Illinois, Champaign-Urbana. He is a distinguished professor and past chairman of the Electrical Engineering Department at the University of California, Los Angeles (UCLA). Before joining UCLA, he was a Senior Research Scientist at NASA Jet Propulsion Laboratory (JPL). He became a Fellow of IEEE in 1985, and was elected as the president of IEEE Antennas and Propagation Society (AP-S) in 1995. Rahmat-Samii has published over 650 journal and conference papers and over 20 books/book chapters in the areas of electromagnetics and antennas. He is the recipient of a large number of awards: two (1992 and 1995) Wheeler Awards for best application papers published in IEEE AP-S Transactions; the University of Illinois ECE Distinguished Alumni Award (1999); the IEEE Third Millennium Medal and the AMTA Distinguished Achievement Award (both in the year 2000); the Technical Excellence Award from JPL (2002); and finally he is the winner of the 2005 International Union of Radio Science (URSI) Booker Gold Medal presented at the URSI General Assembly, New Delhi, India. In 2001, he received an Honorary Doctorate in Physics from one of the oldest Universities in Europe, the University of Santiago de Compostela, Spain. In 2001, he was elected as the Foreign Member of the Royal Academy of Belgium for Science and the Arts. Professor Rahmat-Samii is the designer of IEEE AP-S logo.

Zhan Li was born in Nanjing, China on April 27, 1974. He obtained his B. Sc. and M. Sc Degrees in Electrical Engineering (Radio Electronics) from Nanjing University of Science & Technology in 1995 and 1998, respectively. In March 2005, Zhan received his Ph. D Degree in Electrical Engineering from University of California, Los Angeles (UCLA). He joined the CDMA Research & Development Centre of Nokia Inc. in 2000, where he is currently a Sr. Antenna Design Engineer. His recent research area focuses on the multiple-antenna solution for the handset. He has published 3 journal papers and presented several conference papers at Antenna Propagation Symposia since 2000.

Chapter 4 [Course Lecturers: Prof. Massimo Franceschetti and Prof. Daniele Riccio]

Massimo Franceschetti was born in Napoli, Italy, in 1972, and is currently is Assistant Professor in the Department of Electrical and Computer Engineering of University of California at San Diego (UCSD). He received the Laurea degree, magna cum laude, in Computer Engineering from the University Federico II of Napoli, Italy, in 1997, and the M.S. and Ph.D. degrees in Electrical Engineering from the California Institute of Technology in 1999 and 2003, respectively. Before joining UCSD, he was a post-doctoral scholar at University of California at Berkeley for two years. At Caltech, his doctoral thesis was awarded the C.H. Wilts Prize for best thesis in Electrical Engineering, and the 2000 Walker von Brimer award for outstanding research initiative. Prof. Franceschetti also received (jointly with profs. J. Bruck and L. J. Shulman) in 2004 the S.A Schelkunoff award for the best paper published in IEEE AP-S Transactions, for his work on wave propagation and scattering based on random walk theory. He held visiting positions at the Vrije Universiteit Amsterdam in the Netherlands, the Ecole Polytechnique Federale de Lausanne in Switzerland, and the University of Trento in Italy. His research interests include random networks for communication, wave propagation in random media, and control over networks.

Daniele Riccio was born in Napoli, Italy in 1962. He graduated in Electronic Engineering at the University Federico II of Napoli, Italy, where he is now Professor of Electromagnetics and Remote Sensing. His scientific activity, in the fields of microwave Remote Sensing and radio wave propagation in complex environments, is documented by over 40 papers published on Journals of recognised standard and by 2 books.

Chapter 5 [Course Lecturer: Prof. Mario Gerla]

Mario Gerla received a graduate degree in engineering from the Politecnico di Milano in 1966, and the M.S. and PhD degrees from UCLA in 1970 and 1973. He became IEEE Fellow in 2002. After working for Network Analysis Corporation, New York, from 1973 to 1976, he joined the Faculty of the Computer Science Department at UCLA where he is now Professor. His research interests cover distributed computers, communication systems and wireless networks. He has designed and implemented various network protocols (channel access, clustering, routing and transport) under DARPA and NSF grants. Currently he is leading the ONR MINUTEMAN project at UCLA, with focus on robust, scalable network architectures for unmanned intelligent agents in defense and homeland security scenarios. He is also conducting research on scalable TCP transport for the Next Generation Internet (see www.cs.ucla.edu/NLR for recent publications).

Chapter 6 [Course Lecturer: Prof. Martin Vetterli]

Martin Vetterli was born in Switzerland in 1957. He received his Engineering degree from ETH in Zurich, his MS from Stanford and his Ph.D. from EPFL in Lausanne. In 1986, he joined Columbia University in New York, first with the Center for Telecommunications Research and then with the Department of Electrical Engineering where he was an Associate Professor of Electrical Engineering. In 1993, he joined the University of California at Berkeley, were he was Full Professor until 1997. Since 1995, he is a Professor at EPFL, where he headed the Communication Systems Division (1996/1997) and heads the Audio-visual Communications Laboratory. From 2001 to 2004 he directed the National Competence Center in Research on mobile information and communication systems. He is also a Vice-President for International Affairs at EPFL since October 2004. He has held visiting positions at ETHZ (1990) and Stanford (1998). His research interests are in the areas of applied mathematics, signal processing and communications. He is the co-author of a textbook on "Wavelets and Sub-band Coding", and of over 100 journal papers.

Razvan Cristescu was born in Ploiesti, Romania. He graduated with a PhD in 2004 from EPFL and, after one year as a postdoctoral scholar at Caltech, he is now a Senior R&D Engineer with Becton Dickinson, Sparks MD. He was recipient of the 2005 Chorafas prize for his PhD thesis at EPFL. His scientific activity, in the field of sensor networks, is testified by five papers published on journals of recognised standard.

Chapter 7 [Course Lecturer: Prof. P.R. Kumar]

P. R. Kumar was born in Nagpur, India on April 21, 1952. He obtained his B. Tech. Degree in Electrical Engineering (Electronics) from I. I. T. Madras in 1973, and his M. S. and D. Sc. Degrees in Systems Science and Mathematics from Washington University, St. Louis, in 1975 and 1977, respectively. He was a faculty member in the Department of Mathematics, University of Maryland, Baltimore County, from 1977 to 1984, and since 1985 he has been at the University of Illinois, Urbana-Champaign, where he is currently Franklin W. Woeltge Professor of Electrical and Computer Engineering, and Research Professor of the Coordinated Science Laboratory. He is an IEEE Fellow, was a recipient of the Donald P. Eckman Award of the American Automatic Control Council, and is the 2006 recipient of the IEEE Field Award in Control Systems. He is a co-author of the book "Stochastic Systems," with Pravin Varaiya.

1

Wireless Networks and Their Context

Giorgio Franceschetti and Sabatino Stornelli

1.1 Introduction

This book collects the edited version of the lectures held at a summer school on wireless communications in Capri, Italy, in September of 2004. The aim of the school was to present the communication networks in a wider perspective, thus considering not only their intrinsic message exchange performance, but also taking into account their ability

(i) To collect data and process them;

(ii) To integrate with sensors networks, thus detecting environmental changes and elaborating on them; and

(iii) To finally exert control by implementing far-away actions.

This book concentrates on the scientific and technical aspects of these issues. There is a wide consensus that their practical implementation will result in significant changes in the organization of our society. Accordingly, it seems appropriate to anticipate all the technical matter with a short introduction, aimed at elaborating upon the following points: the impact of these emerging innovations on our life from one side, and the shaping and addressing of the innovations by our society needs on the other. It is obvious that these issues by themselves are worthy of another specific school and may provide the necessary material for another book. Accordingly, the following considerations are very elementary and are limited to the experience of the authors, belonging to the scientific (Giorgio Franceschetti) and industrial (Sabatino Stornelli) communities and not to the sociologist community. But, just for this reason, the following elaborations are derived from an alternative viewpoint and perhaps may be of some interest.

At the moment, the emerging technologies are used to offer innovative services, so that a new market is developing and continuously changing. It is not clear which is the real force that dominates and moves this market. It can be either the technological innovations that drive up and push the offering of innovative services, the demand of new services that pulls the research of new technologies, or probably a combination of these two trends. In addition, new users are adding to the traditional mass market customers: the necessity of rendering our environment safer (*homeland security*) in a very broad sense requires that the public authorities invest in sensing and control by using systems and procedures largely based upon improved and integrated communication networks.

But, irrespective of dominating pushing and pulling forces, the recent technological trends and innovations move the communication network onto a new level, very different from the passive role of connecting remote customers and simply allowing (mainly voice) information exchange. Actually, the network becomes very similar to a biological system that may continuously update and reconfigure itself, depending on the stimuli derived from received messages and from the sensed outside scenario. This network cancels distances, operates in almost real time, and allows people to act in a way similar to the Greek gods: humans become ubiquitous because their actions move at the speed of light; they have immediate access to unlimited past and present data that can be elaborated to predict the future; and they can decide and operate with the power provided by the available full knowledge and by the speed of communication.

There is no doubt that we are moving in this direction. Even if the above statements appear to be a science fiction dream nowadays, they are very likely to become even obsolete in the not too-distant future.

This introductory chapter very simply elaborates on this issue, presenting a number of real-world elementary considerations.

1.2   The Scenario

Information and communication technologies are crucial for the development of countries. The past years have demonstrated that communication technologies are of paramount importance to effectively and efficiently provide new services, by means of large-scale networks deployed over the entire surface of the Earth, thus timely reaching every customer independently of his/her location. Accordingly, the possibility to distribute audio and video information, as well as data, by means of a large-scale communication network opens the way to a wide spectrum of innovative services, characterized by a high level of interactivity with the customer. This represents a strategic opportunity for stimulating demand, growth, and industry development on wireless telecommunications, with the take-off of a new market for multimedia and added value services.

The significant investments made over the years on the ICT (Information and Communication Technology) sector contributed to this technological trend, which is still continuously evolving. The present rise of the TLC (Telecommunication) market, driven by emerging technologies and new opportunities, is pushing the industries to invest in developing multimedia content-based services, accessible at anytime from anywhere, by implementing new available technologies. As an example, Figure 1.1 shows the European TLC market percentage growth. Although this increase has significantly dropped in the years 2001–02 (general recession of global markets and, in Europe, financial demand for Universal Mobile Telecommunication System (UMTS) licenses), the percentage growth trend is rising again. Similar diagrams are available for the United States and Asian countries, with the latter showing a huge increase in very recent years.

New technologies allow new services to be offered, with an increasing appeal to the customer and consequently an increased demand. This, in turn, asks for the improvement of the quality of service (QoS) and customer's satisfaction, thus requiring newer and more innovative technologies. This feed-forward mechanism is still in action today and seems to grow exponentially: technology is moving very fast, and alternative solutions are generated, while others will be offered in the immediate forthcoming years. More important, this technology evolution shows a fundamental impact on the market to provide the *right* service to the *right* user at the *right* place, mixing information and

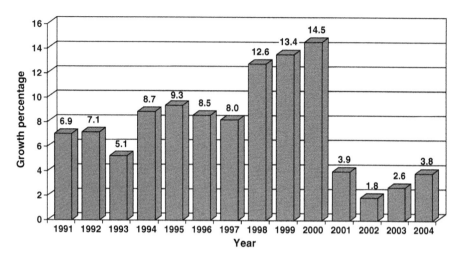

Figure 1.1 Yearly growth of European TLC market, 1991–2004 (source: EITO-IDC 2003).

communications facilities, often combined with information retrieval and processing.

The telecommunication technologies currently offer a lot of products and solutions, both wired and wireless. In particular, wireless communications have received increasing attention during the last few years. The offered solutions are targeted to the mass market (typically with 802.11 standard and mobile 2G/3G [second and third generation] systems), as well as to professional and institutional users, which can also benefit from satellite systems. These solutions are characterized by a common factor: they are *infrastructure*-type networks, devoted to connect mobile users to a fixed network arrangment (Internet or Voice Switched Network), and simply provide single wireless hop communication and/or data exchange. They are pre-deployed, cover a prescribed area, and offer a pre-negotiated service. In other words, the user has to adapt to the network characteristics and is forced to accept the QoS that is offered.

The technological evolution is moving toward a different concept of the network and suggests further evolutions of the 802.11 standard, such as 802.15.4 (ZigBee) or 802.16 (WiMax). As an example, the next mobile generation (4G Mobile or "Beyond 3G") will implement the *convergence* among regional, local, and personal networks, providing a *multiband, multistandard, multimode*, and *multimedia* personal communicator with an embedded broadband wireless core system. The new available technologies will allow the quick set up communications for specific, personalized, short-term applications, covering areas with varying size, where communication infrastructures are not necessarily available, and moving to the new concept of convenient *ad hoc* networking. The new frontier is thus to take into account the dynamic nature of the demand and of the environment, rendering the network able to automatically reconfigure itself. This evolution opens the door to new classes of service providers, that complement the traditional one side-by-side: they differentiate by the offer of specific services, by the segment of customers, and so on. Market niches are created and provide interesting opportunities to gallant entrepreneurs.

Another important issue is the continuous reliability and service assurance, even in emerging situations. This requirement may be guaranteed by an increased integration of the satellite in the communication network. Up to date, a number of telecommunication systems already include the satellite within the communication link to improve and complement the terrestrial one. The satellite improved role is expected to be exploited, especially in the countryside, to provide extensive service access. This is at variance with metropolitan urban areas, wherein the man-made structures (large buildings) generally reduce the sky visibility angle and the satellite access is impaired, thus requiring local wireless service coverage. A viable inclusion of the satellite within the network may be obtained by linking the local cells to a relay point, where a satellite connection can be implemented. This is particularly convenient in emergency situations, to assure the operation of at least a part of the network. It is concluded that pre/during/after disaster needs would rely on different telecommunications systems. In addition, the satellite may play a very significant role with technological improvements of receiving multiple-input multiple-output (MIMO) antennas and in connection with the launch and use of constellations of low orbit satellites.

This dynamic vision of the telecommunication arena nicely complies with our society's increasing information dependence: *anybody* wants access to information *anytime* and *anywhere* (either stationary or on the move): updated reliable news about amenities and entertainment, navigation and traffic, emergencies, etc., must be easily accessible and available on the spot. The trend is from e-business (B2B, B2C, e.g., banking and shopping) and remote collaborative working to a complete *e-activity* era, creating a whole new way of life and culture. It may be defined as the ability and freedom to conduct our activities from anywhere and at anytime.

This evolutionary scenario is also pushing other technologies, such as those relevant to sensors and computing. In fact, significant progress has been made in the area of sensors and devices for detecting audio, video, and other physical parameters of the environment. Indeed, in the next few years we can expect a significant increase in the number of sensors disseminated in the environment to observe it, sense it, and measure its characteristics, collecting a huge amount of data. This information will be processed at the sensor's site, thanks to the decreasing cost and size of processors chips. Therefore, the sensor becomes a *smart* device, whose behavior may change according to the locally processed data and consequent results.

In summary, the overall depicted scenario pushes toward integration of communication, computing, sensing, and control, which was the spirit underlying the school and which is reflected in this book. Forthcoming application will span over a handful of directions: easier access to a wider array of new services by using a single integrated receiving device; environment sensing for control and mitigation of natural disasters, and offer of a safer environment by implementing a pervasive, yet discreet homeland security. The trend is to move along the direction of actively serving the users in an effective and efficient way, rather than just offering them the connection service. This will definitely change our lifestyle and way of thinking.

1.3   The Players

The obvious players within the scenario depicted in Section 1.2 are the service providers and the users, with the latter being further differentiated between institutional and mass market customers. However, this is not quite true, because the technology, too, plays a very significant role; its rapid development continuously suggests or kills possible solutions, thus contributing to the steering of the market offer and demand. Accordingly, the technology itself and its developers, universities, institutional and private research centers, on one side, and manufacturing companies, on the other, are additional important players in the considered scenario. In the following, the interaction between all these players is somehow examined.

A first issue is the evolution of the traditional and the institutional users, with their different needs and requests that can be satisfied by the new technologies under development.

Today, the mass market has gained the possibility to access the advanced technology, previously of almost exclusive use for institutional purposes. During the next few years, sophisticated components embedded on wireless communication devices will transform any environment into a gateway to the communication network. The access will be allowed from any fixed or mobile position. For instance, cars, trains, and airplanes will be equipped to provide office space, where any customer can work with the same (or even better) comfort and facilities available in his/her company office.

New Internet services are appearing, no more related to only data mailing and downloading, but offering a large array of new services spanning from images, videos, texts, books consultation, and e-use to Voice over IP (VoIP) offers. The new border is limited by the new technology that is rapidly emerging and is pushing the existing telecommunication operators to modify their roles since other players are appearing. As a striking example, the VoIP service is fast growing, as shown in Figure 1.2, and it is reasonable to state that within 15 years all communications, including phone calling, will be based on the Internet. The new WiMax technology is a key enabler for exploiting the huge business opportunities that exist in emerging markets. Coming to users as institutional and public authorities, they require the highest available technology level that assures easy and fast deployable solutions, combining a large variety of telecommunication layers with sensor and computing ability to manage emergencies and the associated security aspects. These users generally operate under difficult conditions. They usually centralize the knowledge of the emergency scenario (*situation awareness*), follow up in real-time the evolving

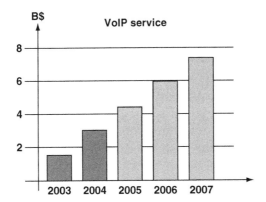

Figure 1.2 Past and projected VoIP technology revenues (B$).

situation, and plan intervention by using a decision making system, which is usually remotely located. These constraints are of paramount importance to the civil protection authorities in the management of natural disasters.

The communication infrastructures utilized by such authorities typically consist of a combination of public and (institutional proprietary) ground (fixed or/or mobile) networks. Public networks suffer high vulnerability and low reliability in cases of natural disasters; their use is limited and their reliability is not adequate, because they are designed for commercial use, thus providing incomplete coverage and/or limited service offer. On the other side, institutional networks (firemen, civil protection, etc.) consist of radio systems operating on dedicated frequency bands and provide communication capabilities on regional and national bases. The increasing demand of security, involving both public authorities and private entities, generates a significant market opportunity to deploy networks aimed at these specific services.

Communications, precise position, and integrity/reliability represent the real requirements for the institutional market to be captured. In most cases, tracking and alarming devices are exploited by public organizations. Private companies could be part of the framework, providing services and products to perform pinpointing services. The operative scenario is different compared to a few years ago; a clear understanding and precise modelling of the environment and highly reliable systems both in terms of communication and sensor systems are imperative. A secure environment starts from the prevention phase that creates a sense of safety for citizens, with consequent improvement of their quality of life.

A second issue is the interaction between product manufacturers and service providers within the considered scenario.

Up to now we have experienced a massive offer of emerging technologies, with an attempt to propose specific products. However, such an offer will not guarantee to automatically create innovative services if the impact of some constraints onto the innovation process is not taken into account. This requires to elucidate the interactions among the quoted actors, product manufacturers and service providers.

Product manufacturers and service providers generally move only to *force* their products rather than to take a holistic view of the scenario to whom the emerging proposed technologies should be addressed. Service providers concentrate their effort on mass market, investing on existing or emerging technologies (for instance, the forthcoming wireless WiMax system). They face the challenges related to mass market needs and struggle to remain profitable within the higher level of (institutional) customer expectations. At the same time, the suppliers, i.e., the manufacturers, must find innovative ways to deliver value to their products, otherwise they may also risk going out of business. Accordingly, they explore the new wireless communication technologies particularly appropriate to institutional users. Both technologies, devoted to mass market and institutional applications, must obviously comply with regulatory aspects, in particular the use of carrier frequencies and associate bandwidths, whose boundaries are not necessarily precisely enforced. It follows that technology availability does not necessarily guarantee its fruitful exploitation. In the absence of clear regulatory rules, manufacturers could re-address their business toward mass market products, giving up technology exploitation for institutional users, with evident negative long-range impact for the citizens community. Therefore, the communications industry needs to identify *new* rules for introducing *new* operator figures able to look at the *new* customers' *new* needs.

Providers and suppliers both operate in a very complex market environment, driven by often contrasting forces and very different regulatory settings. Therefore, the business strategies must change accordingly. This requires a great effort in terms of devoting significant human resources to analyze, model, and explore the entire market value chain, including all the industrial companies as well as all species of existing and expected possible customers. This is at variance with the exploited approaches to competition during even the most recent years: the emerging technological framework requires to develop new business models for mass, professional, and institutional market applications. But this convergence of a significant number of specific technologies, together with their deep knowledge, does not automatically imply a system's integration ability.

In view of the above considerations, it is reasonable to think of a cooperative value chain scenario, where all actors are involved. Each actor of the chain is a provider and a user at the same time that offers and requests his/her specific products and needs among hardware components, software platforms, market expectations, regulatory issues information, and so on. Besides service providers and manufacturers, customers and regulating authorities are also included in the chain. The former may contribute to the network operation (for instance, a local area multi-hop network), while the latter may contribute by shaping up the regulatory issues upon the changing technological scenario.

The presented integrated business scenario requires the implementation of a joint-enterprise model to identify and isolate the component operational functions, so that the role of the involved actors in the value chain is preserved. This allows them to invest in a more specific activity in the frame of the overall cooperation. They should possess an outstanding experience in the addressed business area, thus significantly contributing to the overall business according to their basic economic interests. This approach eventually leads to an extended service provider entity, a form of a consortium and an integrated enterprise that spans from users, through manufacturers, to services, applications, and content providers.

In conclusion, the economic driver is to identify how to take part in this new highly changeable development, fostering technological, market, regulatory, and policy convergence.

An interesting example in response to this challenge in Italy may be quoted by the second author of this chapter: Finmeccanica, a state owned industrial holding, is reorganizing its companies, Selex-SI, Selex-Com, Selex-SeMa, Elsag, Telespazio, Seicos, and others, in order to cover the entire value chain of the sector by using the specific competences existing in the group, and to continuously pursue emerging technology and innovative services within an organized fashion. In such a way the group provides extensive experiences and assets for the satisfaction of a broad range of customer needs and desired solutions. Similar experiences are either existing or materializing in all other industrial countries.

To summarize, the scenario requires a high level of flexibility and adaptability, changing the roles to the full satisfaction of the customer, even though the overlap among markets, technologies, policies, rules, and so on may generate huge dilemmas to decide the right direction to take and to avoid collisions. The latter statement brings to mind the following questions. Which are the resulting impacts and what can be done to minimize them? What are the boundaries of the regulatory issues and how can the most appropriate direction be shared? The right answer seems to be setting up a plan according to the country strategy about regulations, site access, and incentives in order to choose the appropriate solution to the quoted problems.

A third issue is related to university and research center players. It is well known that the usual research and development policy of industrial companies is mostly steered to the near (and perhaps medium) range applications. In addition, appropriate activity and investments in industrial R&D should design technologies starting from services and boundary constraints. Examples of the latter are frequency regulatory plan and site accessibility. This is crucial for the stimulation of entrepreneurship. This is certainly understandable and usually convenient, but not sufficient in the very rapidly changing scenario of communication technology. This deficiency may be corrected by establishing and/or improving intensive relations with universities and research centers, which can provide vision and explore the long-term evolution of telecommunication science and technology. It is also noted that new technology-based firms, usually in the form of small start-ups, cannot be financially self-sustaining in the long term; their valuable expertise should be financially nourished and exploited. Finally, the R&D process is not concluded with the technology production. It should learn from the earned operational experience by using the exploited technology. In other words, sometimes the reality overcomes some R&D results.

The previously presented joint-enterprise model can achieve further success if universities and research centers are deeply included in the scenario. They should be strongly involved in the innovation process, providing, from their unbiased view, innovative ideas and brilliant solutions. The telecommunication sector and, in particular, the wireless one could gain huge advantages through such cooperation, capitalizing the extraordinary already obtained results by long-term research investment.

1.4   Concluding Remarks

At the end of previous considerations, it would be desirable to present a possible scenario of the future: where are we going? This is not possible for at least two reasons. First, the material that has been presented is too simple and qualitative. Quantitative information would certainly be needed. But, even in the presence of this additional information, predicting the future of a situation whose time derivative is very high and randomly oscillating is an ill-posed problem (from the mathematical viewpoint). Its solution is hard or even impossible to obtain. In spite of this, we believe that some few points might be assessed.

As far as the operational scenario is concerned, we believe that all the above statements, scattered under previous sections, are valid: the integration process between communication, computing, sensing, and control will materialize; new actors will enter the scene, as well as different species of new applications and entrepreneurs; different companies of complementary expertise and mission will gather in alliances; and long-term vision and investment will be provided at a certain extent.

As additional comment, the success of a new technology will be not only dependent on its value and timely appearance, but politics related to institutional organizations will play their role. Furthermore, the choice of new solutions does not necessarily imply the end of the previous one. The latter may be very specific (e.g., TETRA [Terrestrial Trunked Radio] used by police forces in UK, Germany and Italy) or can be revisited in light of additional technological improvements (ADSL [Asymmetric Digital Subscriber Line] is an example).

Finally, it will be realized very soon that a novel type of scientist is necessary to master all the areas of the telecommunication sector. The design of new, sophisticated networking devices and protocol architectures requires a great deal of *cross-layer* optimization. For example, the video encoder used to deliver video on demand services must be *aware* of the radio characteristics (modulation, channel propagation, jamming, motion) of the wireless links along the path. Moreover, encoder and radio designs must be adaptively adjusted to user demands. This is far away from the olden days when each architectural layer was independently designed. This new, cross layer, interdisciplinary design paradigm requires not only in-depth expertise in one specific layer, but also a well rounded scientific background that integrates physics, applied mathematics, electromagnetics, communication, controls, computer and information sciences, coupled with an appreciation for social and behavioral science values, problems, and models. The above requirements would imply a very intensive study and dedication, but this must not happen at the expense of the vision qualities of the student. This is not an easy task. It is the responsibility of the academia to realize a program that can educate and form the *telecommunications engineers* of the future.

2

The Wireless Communications Physical Layer

Michael P. Fitz, Cong Shen, and Michael Samuel

2.1 Historical Perspectives

The subject of wireless communications is about getting information from point A to point B using radio waves. Born two centuries ago, this field continues today in a vast number of applications. Since prehistoric times, humans have needed communications for the building of wealth and the waging of war. The aid of government-sponsored monopolies and these social forces have continuously advanced communications performance. It is perhaps interesting to note that the first electronic communications (telegraphy) sent digital data (words were turned into a series of electronic dots and dashes). World War I led to great advances in wireless technology, and television and radio broadcasting soon followed. In these applications the transmitted information sources were analog. The digital revolution was spawned by the need for the telephone network to multiplex and automatically switch a variety of phone calls. A further technology boost was given during World War II in wireless communications and system theory. The Cold War led to rapid advances in satellite communications and system theory as the race for space gripped the world's major technology innovation centers. The invention of the semiconductor transistor and the march of Moore's law have spurred the progression of innovation since the early 1980s. The evolving power of the microprocessor, the embedded computer, and the signal processor has enabled algorithms that were considered preposterous at their formulation to see cost-effective implementation. Distilling this 150 plus years of innovation into a short chapter is a challenge, but one these authors arrogantly attempt.

The relative growth rate of electronic communications is phenomenal. The world community has gone in a very short period of time from accepting message delivery delays of weeks down to seconds. From the invention of the microphone, the electric motor, the electronic tube, and the transistor up to the laser, engineers and physicists have made great technology leaps forward. These technological leaps have made great advances in communications possible. As technology has advanced, the communication engineering profession has become multifaceted and specialized over time. What once was a field where non-experts could contribute[*] prior to 1900 became a field where great specialization was needed in the post-1900 era. Two areas of specialization formed through the 1900s: the devices engineer and the systems engineer. The devices engineer focuses on designing technology to complete certain tasks. Devices engineers, for example, build antennas and oscillators and are heavily involved with current technology. Systems engineers try to put devices together in a way that will work as a system to achieve an overall goal. System engineers try to form mathematical models for how systems operate and use these models to design and specify systems. This chapter is written with a **systems engineering** perspective. Systems engineering in communications did not come to be a formalized field until the early 1900s; hence none of the references in this chapter were published before 1900. Some interesting historical system engineering references are references 3, 12, 25, 35, 45, 48 and 62.

2.2 Digital Communication Basics

Point-to-point binary data communications is the main theme of this chapter. The system model for such a communication system is given in Figure 2.1. A source of binary encoded data is present, and it is desired to transmit this data to a binary data sink across a physical channel. The data output by the source is represented by a $K_b \times 1$

---

[*] For example, Samuel Morse (of Morse code and telegraph systems fame in the United States) was a professor in the liberal arts.

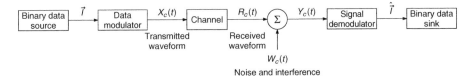

Figure 2.1   A model for point-to-point data communication.

dimensional vector, $\vec{I}$, whose components take values 0,1. This vector is then mapped into one of 2^{K_b} analog waveforms represented by the waveform $X_c(t)$. The transmitted waveform is put through a channel of some sort and corrupted by noise or interference. The composite received signal, $Y_c(t)$, is then used to estimate which one of the possible vectors led to the transmitted waveform. This estimate is denoted $\hat{\vec{I}}$, and this estimate is passed to the data sink.

2.2.1   Complex Baseband Representation of Bandpass Signals

All wireless communication systems operate by modulating an information bearing waveform onto a sinusoidal carrier.[**] However, one should notice that the carrier frequency of the transmitted signal is not the component which contains the information. Instead, it is the signal modulated on the carrier which carries the information. Hence a method of characterizing a communication signal which is independent of the carrier frequency is desired. This has led communication system engineers to use a **complex baseband representation** of communication signals to simplify their job. Nearly all of the modern communication systems can be and typically are analyzed with this complex baseband representation. This chapter highlights the complex baseband representation for deterministic signals. Other references that develop these topics in more detail are references 7, 27, 42 and 43. One advantage of the complex baseband representation is simplicity. All signals are lowpass, and the fundamental ideas behind modulation and communication signal processing are easily developed. Additionally, any receiver that processes the received waveform digitally uses the complex baseband representation to develop the baseband processing algorithms. In fact, complex

---

[**]One might argue that ultra wideband radio (UWB) is an exception to this rule, but even a vast majority of UWB radios have a sinusoid carrier.

baseband representation is so prevalent in engineering systems that the most widely used tool, *Matlab*, has been configured by default to process all variables in a program as complex signals.

Definition 2.1   *A bandpass signal, $x_c(t)$, is a signal whose one-sided energy spectrum is both (1) centered at a non-zero frequency, f_C, and (2) does not extend to zero frequency.*

The two-sided transmission bandwidth of a signal is typically denoted by B_T Hertz so that the one-sided spectrum of the bandpass signal is zero except in $[f_C - B_T/2, f_C + B_T/2]$. This implies that a bandpass signal satisfies the following constraint: $B_T/2 < f_C$. Since a bandpass signal, $x_c(t)$, is a physically realizable signal, it is real valued, and consequently the energy spectrum will always be even symmetric around $f = 0$. The relative sizes of B_T and f_C are not important, only that the spectrum takes negligible values around zero frequency.

A bandpass signal has a representation of

$$x_c(t) = x_I(t)\sqrt{2}\cos(2\pi f_c t) - x_Q(t)\sqrt{2}\sin(2\pi f_c t) \qquad (2.1)$$

$$= x_A(t)\sqrt{2}\cos(2\pi f_c t + x_P(t)), \qquad (2.2)$$

where f_c denotes the carrier frequency with $f_C - B_T/2 \leq f_c \leq f_C + B_T/2$. The signal $x_I(t)$ in (2.1) is normally referred to as the **in-phase (I)** component of the signal, and the signal $x_Q(t)$ is normally referred to as the **quadrature (Q)** component of the bandpass signal. $x_I(t)$ and $x_Q(t)$ are real-valued lowpass signals with a one-sided non-negligible energy spectrum no larger than B_T Hertz. Two items should be noted:

• The center frequency of the bandpass signal, f_C, and the carrier frequency, f_c, are not always the same.

• The $\sqrt{2}$ term is included in the definition of the bandpass signal to ensure that the bandpass signal and the baseband signal have the same power/energy.

The carrier signal is normally thought of as the cosine term; hence the *I* component is in-phase with the carrier. Likewise, the sine term is 90° out-of-phase (in quadrature) with the cosine or carrier term; hence the *Q* component is quadrature to the carrier. Equation (2.1) is known as the *canonical form* of a bandpass signal. Equation (2.2) is the amplitude and phase form of the bandpass signal, where $x_A(t)$ is the **amplitude** of the signal, and $x_P(t)$ is the **phase** of the signal. A bandpass signal has two degrees of freedom and the I/Q or the amplitude and phase representations are equivalent. The transformations between the two representations are given by

$$x_A(t) = \sqrt{x_I^2(t) + x_Q^2(t)} \qquad\qquad x_P(t) = \tan^{-1}\left[x_Q(t), x_I(t)\right] \qquad (2.3)$$

and

$$x_I(t) = x_A(t)\cos\left(x_P(t)\right) \qquad\qquad x_Q(t) = x_A(t)\sin\left(x_P(t)\right). \qquad (2.4)$$

The particulars of the communication design analysis determine which form for the bandpass signal is most applicable.

A complex-valued signal, denoted the **complex envelope**, is defined as

$$x_z(t) = x_I(t) + jx_Q(t) = x_A(t)\exp\left[jx_P(t)\right]. \qquad (2.5)$$

The original bandpass signal can be obtained from the complex envelope by

$$x_c(t) = \sqrt{2}\Re\left[x_z(t)\exp\left[j2\pi f_c t\right]\right], \qquad (2.6)$$

where $\Re[\cdot]$ denotes taking the real part of a complex number. Since the complex exponential only determines the carrier frequency, the complex signal, $x_z(t)$, contains all the information in $x_c(t)$. Using this complex baseband representation of bandpass signals greatly simplifies the notation for communication system analysis.

The next item to consider is methods to translate between a bandpass signal and a complex envelope signal. Basically, a bandpass signal is generated from its I and Q components in a straightforward fashion corresponding to (2.1). Likewise, a complex envelope signal is generated from the bandpass signal with a similar architecture. The idea behind bandpass to baseband down-conversion can be understood by using trigonometric identities to give

$$x_1(t) = x_c(t)\sqrt{2}\cos(2\pi f_c t) = x_I(t) + x_I(t)\cos(4\pi f_c t) - x_Q(t)\sin(4\pi f_c t)$$
$$x_2(t) = x_c(t)\sqrt{2}\sin(2\pi f_c t) = -x_Q(t) + x_Q(t)\cos(4\pi f_c t) + x_I(t)\sin(4\pi f_c t).$$

$$(2.7)$$

In Figure 2.2 the lowpass filters (LPF) remove the $2f_c$ terms in (2.7). Note in Figure 2.2 that the boxes with $\pi/2$ are phase shifters

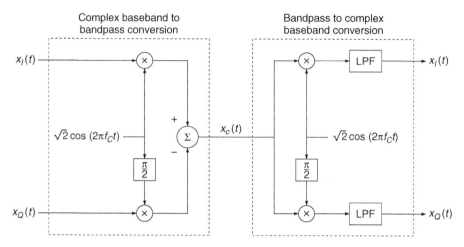

Figure 2.2 Schemes for converting between complex baseband and bandpass representations. Note that the LPF simply removes the double frequency term associated with the down conversion.

(i.e., $\cos(\theta - \pi/2) = \sin(\theta)$) typically implemented with delay elements. The structure in Figure 2.2 is fundamental to the study of all carrier modulation techniques.

2.2.2 Digital Transmission

Digital communication systems have a modulator and demodulator. The modulator produces an analog signal that depends on the digital data to be communicated. This analog signal is transmitted over a channel (radio propagation). The demodulator takes the received signal and constructs an estimate of the transmitted digital data.

Definition 2.2 *Digital modulation is a transformation of $\vec{I}$ into a complex envelope, $X_z(t) = \Gamma_m(\vec{I})$.*

It should be emphasized that digital communication is achieved by producing and transmitting analog waveforms. There is no situation where communication takes place that this transformation from digital to analog does not occur. Similarly, it should be emphasized that the modulator must be capable of generating 2^{K_b} continuous time waveforms to represent each of the possible binary data vectors produced by the binary data source. When necessary, the possible data vector will be enumerated as $\vec{I} = i$, $i \in \{0, \ldots, 2^{K_b} - 1\}$, and the possible transmitted waveforms will be enumerated with $X_z(t) = x_i(t)$. For the sake of discussion, the transmitted data vector will be modeled as random, and the probability that each word will be chosen to be transmitted is denoted $\pi_i = P(\vec{I} = i)$.

Definition 2.3 *Digital demodulation is a transformation of $Y_c(t)$ into estimates of the transmitted bits, $\hat{\vec{I}}$.*

Demodulation takes the received signal, $Y_c(t)$, and down-converts to the baseband signal, $Y_z(t)$. The baseband signal is then processed to produce an estimate of the transmitted data vector $\vec{I}$. This estimate will be denoted $\hat{\vec{I}}$. Demodulation is the process of producing an $\hat{\vec{I}}$ from $Y_z(t)$ via a function $\Gamma_d(Y_z(t))$. It is worth noting at this point that the word modem is actually an engineering acronym for a device that is both a *mo*dulator and a *dem*odulator. The term modem has become part of the English language and is now synonymous with any device that is used to transmit digital data (computer modem, cable modem, wireless modem, etc.).

2.2.3   Performance Metrics for Digital Communication

In evaluating the efficacy of various designs, the performance metrics commonly applied in engineering design must be examined. The most commonly used metrics for digital communications are

- Fidelity – This metric typically measures how often data transmission errors are made given the amount of transmitted power.

- Complexity – This metric almost always translates directly into cost.

- Spectral Efficiency – This metric measures how much bandwidth a modulation uses to implement the communication.

2.2.3.1   Fidelity
Fidelity in digital communication is reflected by how often transmission errors occur as a function of the signal-to-noise ratio (SNR). Transmission errors can be either bit errors (one bit in error) or frame errors (any error in a message or packet). The application often determines the appropriate error metric. With data transmission at a fixed

transmit power, P_{x_c}, the reliability of any data communication can be increased by lowering the speed of the data communication. A lower speed transmission implies that the receiver bandwidth will be smaller, and consequently the SNR can be made higher. In data communication, a transmission rate fair measure of SNR is the ratio of the average received energy per bit E_b over the one-sided noise spectral density N_0, E_b/N_0. Parameterizing fidelity by E_b/N_0 is widely used in communication theory and has become the industry standard.

2.2.3.2 Complexity

Complexity is a quantity that requires engineering judgment to estimate. The cost of a certain level of complexity has changed over time. A good example of this tradeoff changing over a short period of time was seen in the land mobile telephony market in the 1990's. Early in the decade many people resisted a move to a standard based on Code Division Multiple Access (CDMA) technologies based on the cost and complexity of the handheld phones. By the end of the decade the proposed telecommunication standards had become much more complex, but the advances in circuit technology allowed low cost implementations. It is worth noting that some of the processor chips of modern mobile phones have even more transistors than modern personal computer processor chips.

2.2.3.3 Spectral Efficiency

The spectral efficiency of a communication system is typically a measure of how well the system is using the bandwidth resource. In this chapter the bit rate of communication is denoted W_b bits per second and the transmission bandwidth is denoted B_T. Bandwidth costs money to acquire, and the owners of this bandwidth want to communicate at as high a data rate as possible. Examples are licenses to broadcast radio signals or the installation of copper wires to connect two points. Hence spectral efficiency is very important for people who try to make money selling communication resources. For instance, twice the spectral efficiency implies twice the revenue to wireless service providers. The precise measure of spectral efficiency that we will use in this chapter is defined as

$$\eta_B = \frac{W_b}{B_T} \quad \text{bits/second/Hz}.$$

The goal of this section is to associate a spectral characteristic or a signal bandwidth with a digital modulation. This spectral characteristic determines the bandwidth of a radio that needs to be designed to support the transmission as well as the spectral efficiency of a digital transmission scheme. The way this will be done is to note that if the data being transmitted are known, the transmitted signal, $x_z(t)$, is a deterministic energy signal. The spectral characterization of deterministic energy signals is given by the energy spectrum

$$G_{x_z}(f) = |X_z(f)|^2 = |\mathcal{F}\{x_z(t)\}|^2, \tag{2.8}$$

where $\mathcal{F}\{\cdot\}$ denotes the Fourier transform. The function that will be used throughout this chapter to describe the spectral characteristics of a transmitted signal is the average energy spectrum per bit.

Definition 2.4   *The average energy spectrum per bit for a transmitted signal, $X_z(t)$, where K_b bits are transmitted is*

$$D_{X_z}(f) = \frac{E\left[G_{X_z}(f)\right]}{K_b}. \tag{2.9}$$

It should be noted that the expectation or average in (2.9) is over the random transmitted data bits. Since the random transmitted bits are discrete random variables, the expectation will be a summation versus the probability of each of the possible transmitted words. Consequently, the general form for the average energy spectrum per bit is

$$D_{X_z}(f) = \sum_{i=0}^{2^{K_b}-1} \pi_i G_{x_i}(f). \tag{2.10}$$

The transmission bandwidth, B_T, of a digital communications signal can be obtained from the average energy spectrum per bit, $D_{X_z}(f)$.

2.2.3.4   Other Important Characteristics

Many times in communication applications other issues besides spectral efficiency, complexity, and fidelity are important. For example, the size, weight, and battery usage of a handheld mobile device are important for the user. Often in wireless communications energy efficiency of the algorithms are of paramount importance. To increase the talk time of a mobile phone, an algorithm will sometimes give up performance to use less energy. One important issue in mobile devices is the linearity of the final amplifier before the antenna. A high power linear amplifier is both expensive and consumes larger amounts of current, so this is not a desirable characteristic for a mobile device. On the other hand, high power is often needed to communicate with a remote base station or a satellite. Likewise, non-linear amplifiers, which are more energy efficient, often produce unacceptable distortion or spectral regrowth. In many handheld devices, modulations are chosen to minimize the requirement on the linearity of the power amplifier.

2.2.4   Some Limits on Performance of Digital Communication Systems

Digital communications is a relatively unique field in engineering in that there is a theory that gives some performance limits for data transmission. The body of work that provides us with these fundamental limits is information theory, and the founder of information theory was Claude Shannon.[48] While this chapter cannot derive all the important results from information theory, it will attempt to highlight those results related to digital communications. Interested readers can refer to[17] for more details.

One of Shannon's important contributions was to identify that every channel had an associated capacity, C, and reliable (arbitrarily small error probability) transmission is possible on the channel when $W_b < C$. A channel of significant interest is the channel which experiences an Additive White Gaussian Noise (AWGN) distortion. For this AWGN channel when the signal uses a transmission bandwidth of B_T, Shannon identified the capacity as[48]

$$C = B_T \log_2 (1 + SNR), \qquad (2.11)$$

where $SNR = P_s/P_N$, P_s is the signal power, and P_N is the noise power. This immediately leads to a constraint on the spectral efficiency that can be reliably achieved

$$\eta_B < \log_2 (1 + SNR). \qquad (2.12)$$

Equation (2.12) unfortunately states that to achieve a linear increase in spectral efficiency a communication engineer must provide exponentially greater received SNR. Hence in most communication system applications the spectral efficiencies achieved are usually less than 15 bits/s/Hz (often much less).[†]

Further insight into the problem is gained by reformulating (2.12). By using

$$P_N = N_0 B_T \qquad P_s = E_b W_b, \tag{2.13}$$

it is easy to obtain

$$\eta_B < \log_2 \left(1 + \frac{E_b}{N_0} \frac{W_b}{B_T} \right) = \log_2 \left(1 + \frac{E_b}{N_0} \eta_B \right). \tag{2.14}$$

The achievable spectral efficiency versus E_b/N_0 is represented in Figure 2.10 (the upper bound). The line in Figure 2.10 represents the solutions to the equation

$$\eta_B = \log_2 \left(1 + \frac{E_b}{N_0} \eta_B \right). \tag{2.15}$$

For a given E_b/N_0, Shannon proved that reliable communication at spectral efficiencies below the line in Figure 2.10 is achievable, while spectral efficiencies above the line are not achievable. Throughout the remainder of the chapter the goal will be to give an exposition on how to design communication systems that have operating points which can approach this ultimate performance given in Figure 2.10 at a reasonable complexity.

The results in Figure 2.10 provide some interesting insights for how communication systems should be designed. In situations where bandwidth is the most restricted resource, the goal then is to drive the received E_b/N_0 to as large a value as possible. For example, many telecommunication systems have designed operating points where $E_b/N_0 > 10$ dB. In situations where E_b is the most restricted resource, it is possible to still achieve reliable communication by reducing the spectral efficiency. Communications with deep space probes are limited by the amount of power that can be received. Communication systems for deep space communication are designed most often to have relatively low bit rates and by setting $\eta_B < 1$. Figure 2.10 shows that there is a limit on how small E_b/N_0 can be made and still maintain reliable communications. This minimum is

---

[†] In the past decade researchers have found that using multiple antennas at both transmitter and receiver sides can greatly increase the spectral efficiency of wireless communications. This will be discussed in more detail in the sequel.

$E_b/N_0 = \ln 2 = -1.59$ dB. These results from information theory provide benchmarks to calibrate performance as this chapter investigates the theory of wireless digital communications.

2.2.5 Optimum Demodulation

Now we want to consider the digital communication system design problem of transmitting K_b bits. The bit sequence is denoted $\vec{I} = [I(1)\ I(2)\ \ldots\ I(K_b)]^T$, where $I(k)$ take values 0,1. For simplicity of notation, a particular sequence of bits can be designated by the numeric value the bits represent, i.e., $\vec{I} = i$, $i \in \{0, \ldots, M-1\}$, where $M = 2^{K_b}$. To represent the M values the bit sequence can take, M different analog waveforms should be available for transmission. Denote by $x_i(t)$, $i \in \{0, \ldots, M-1\}$ the waveform transmitted when $\vec{I} = i$ is to be transmitted. Here again we will assume the analog waveforms have support on $t \in [0, T_p]$. It is worth noting that this problem formulation gives a bit rate of $W_b = K_b/T_p$ bits per second.

2.2.5.1 General Demodulators
A first receiver to be considered is the maximum *a posteriori* word demodulator (MAPWD). This MAPWD can again be shown to be the minimum word error probability receiver using Bayes detection theory. Statistical decision theory[41] leads to

$$\hat{\vec{I}} = \arg \max_{i \in \{0, \ldots, M-1\}} P\left(\vec{I} = i | y_z(t)\right), \tag{2.16}$$

where the arg max notation refers to the particular M-ary word that has the maximum *a posteriori probability*. Using the results of Poor[41] and Van Trees[57] this MAP decoding rule becomes

$$\hat{\vec{I}} = \arg \max_{i \in \{0, \ldots, M-1\}} \exp\left[\frac{2}{N_0}\Re[V_i(T_p)] - \frac{E_i}{N_0}\right]\pi_i, \tag{2.17}$$

where

$$V_i(t) = \int_{-\infty}^{\infty} y_z(\tau)x_i^*(T_p - t + \tau)d\tau \tag{2.18}$$

is denoted the ith matched filter output, and

$$E_i = \int_{-\infty}^{\infty} |x_i(t)|^2\, dt \tag{2.19}$$

is the energy of the ith analog waveform. The ith matched filter output when sampled at $t = T_p$ again will give a correlation of the received signal with the ith possible transmitted signal.

The demodulator in the case of equal priors, i.e., $\pi_i = 1/M$, $\forall i$, can be greatly simplified. We will denote this demodulator as the maximum likelihood word demodulator (MLWD). Since all terms have an equal π_i, this common term can be cancelled from each term in the decision rule, and since the $\ln(\cdot)$ function is monotonic, the MLWD is given as

$$\hat{\hat{I}} = \arg \max_{i \in \{0, \ldots, M-1\}} \Re\left[V_i(T_p)\right] - \frac{E_i}{2}. \tag{2.20}$$

Decoding is accomplished by selecting the binary word associated with the largest matched filter output and energy correction.

The important thing to notice for both MAPWD and MLWD is that the optimal demodulator complexity increases exponentially with the number of bits transmitted. The number of matched filters required in each demodulator is $M = 2^{K_b}$. Consequently, the complexity of these demodulation schemes is $O(2^{K_b})$. The notation $O(x)$ implies that the complexity of the algorithm is proportional to x, i.e., a constant times x. This complexity is obviously unacceptable if large files of data are to be transmitted. To make data communications practical, ways will have to be developed that make the complexity linear in the number of bits sent, i.e., $O(K_b)$.

In the following two obvious and important examples of M-ary carrier modulated digital communication are considered: M-ary frequency shift keying (MFSK) and M-ary phase shift keying (MPSK).

2.2.5.2   M-ary FSK

MFSK modulation sends the word of information by transmitting a carrier pulse of one of M frequencies. This is an obvious simple signalling scheme, and one used in many early modems. The signal set is given as

$$x_i(t) = \begin{cases} \sqrt{\dfrac{K_b E_b}{T_p}} \exp\left[j2\pi f_d(2i - M + 1)t\right] & 0 \le t \le T_p, \\ 0 & \text{elsewhere} \end{cases} \tag{2.21}$$

where f_d is known as the frequency deviation. The frequency difference between adjacent frequency pulses in the signal set is $2f_d$. It is apparent that each waveform in an MFSK signal set has equal energy that has here been set to $E_s = K_b E_b$.

The average energy spectrum per bit is again used to characterize the spectral efficiency. Recall that the average energy spectral density per bit is given for M-ary modulations as

$$D_{x_z}(f) = \frac{1}{K_b} \sum_{i=0}^{M-1} \pi_i G_{x_i}(f). \tag{2.22}$$

The energy spectrum of the individual waveforms is given as

$$G_{x_i}(f) = K_b E_b T_p \mathrm{sinc}^2((f - f_d(2i - M + 1))T_p), \tag{2.23}$$

where $\mathrm{sinc}(x) = \sin(\pi x)/(\pi x)$. We can prove that the minimum frequency separation needed to achieve an orthogonal modulation is $f_d T_p = 0.25$ and that by considering (2.22) and (2.23) it is obvious that the spectral content is growing proportional to $B_T = f_d(2^{K_b+1})$. An example of each of the individual energy spectrums (dotted lines) and the average energy spectrum (solid line) for 8FSK is plotted in Figure 2.3. The transmission rate of MFSK is $W_b = K_b/T_p$. The spectral efficiency then is approximately $\eta_B = K_b/2^{K_b-1}$ and decreases with the number of bits transmitted or equivalently decreases with M. Conversely, MFSK provides monotonically increasing fidelity with M[42] and thus has found use in practice when lots of bandwidth is available and good fidelity is required.

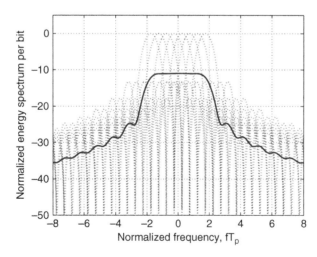

Figure 2.3   For 8FSK the $G_{x_i}(f)$ for $i \in \{0, \ldots, 7\}$ and $D_{x_z}(f)$.

The advantages of MFSK are summarized as

- MFSK is very simple to generate: simply gate one of M oscillators on, depending on the word to be sent.

- Fidelity improves monotonically with K_b. This is counter intuition as one might expect performance degrading with increasing K_b.

The disadvantages of MFSK are summarized as

- The bandwidth increases exponentially with K_b, and hence the spectral efficiency decreases with K_b.

- Complexity increases exponentially with K_b.

2.2.5.3   *M*-ary PSK

MPSK modulation sends the word of information by transmitting a carrier pulse of one of M phases. This modulation is also used in many modems. The first form of the signal set one might consider is given as

$$x_i(t) = \begin{cases} \sqrt{\dfrac{K_b E_b}{T_p}} \exp\left[j\dfrac{\pi(2i+1)}{M} \right] & 0 \le t \le T_p \\ 0 & \text{elsewhere.} \end{cases} \quad (2.24)$$

It is apparent that each waveform in an MPSK signal set has equal energy that has here been set to $E_s = K_b E_b$. The phases have been chosen uniformly spaced around the unit circle.

The average energy spectrum per bit is again used to characterize the spectral efficiency of MPSK. Recall that the average energy spectral density per bit is given for M-ary modulations as

$$D_{x_z}(f) = \frac{1}{K_b} \sum_{i=0}^{M-1} \pi_i G_{x_i}(f). \quad (2.25)$$

Recall that the energy spectrum of the individual waveforms is given as

$$G_{x_i}(f) = E_b T_p \left| \exp\left[j\frac{\pi(2i+1)}{M} \right] \right|^2 \text{sinc}^2\left(f T_p\right) = E_b T_p \text{sinc}^2\left(f T_p\right). \quad (2.26)$$

The occupied bandwidth of MPSK, $B_T \propto 1/T_p$, does not increase with M, while the bit rate does increase with M, $W_b = K_b/T_p$, to provide a spectral efficiency of $\eta_B = K_b$. Unfortunately, the fidelity of MPSK modulation decreases with increasing M.[42] Consequently, MPSK modulations are of interest in practice when the available bandwidth is small and the SNR is large. It should also be noted that the optimum demodulator of MPSK only needs one matched filter and one decision device, and thus has a significant advantage in complexity.

The advantages of MPSK are summarized as

- MPSK is very simple to generate: simply change the phase of an oscillator to one of the M values, depending on the word to be sent.

- There is no increase in the bandwidth occupancy with increasing K_b. Consequently, spectral efficiency increases with K_b.

- Demodulation complexity does not increase exponentially with K_b.

The disadvantages of MPSK are summarized as

- Fidelity decreases monotonically with K_b.

2.2.6   Discussion

This section introduces the concept of digital modulation and demodulation and two example modulations to transmit K_b bits of information: MFSK and MPSK. MPSK has the advantage of being able to supply an increasing spectral efficiency with K_b at the cost of requiring more E_b/N_0 to achieve the same performance. MFSK can provide improved performance with K_b, but at a cost of a loss of spectral efficiency. Additionally, the decoding complexity of MPSK is significantly less than the decoding complexity of MFSK. As a final point it is worth comparing the spectral efficiency performance of these two modulations with the upper bounds provided by information theory (see Section 2.2.4). We will denote reliable communication as being an error rate of 10^{-5}. The operating points of MFSK and MPSK and the upper bound on the possible performance are plotted in Figure 2.10. It is clear from this graph that different modulations give us a different set of points in a performance versus spectral efficiency tradeoff. Also the two examples considered in this section have a performance much lower than the upper bound provided by information theory. This is still not too disturbing as lots of digital communication theory is left to explore.

2.3 Orthogonal Modulations

Orthogonal modulations, such as Orthogonal Frequency Division Multiplexing (OFDM), have been the primary vehicle for cost-effective data communications. In this section we will introduce the three most widely used orthogonal modulations and overview their engineering utility.

2.3.1 Orthogonal Frequency Division Multiplexing

A commonly used modulation that admits a simple optimal bit demodulation is OFDM.[15] OFDM has found utility in telephone, cable, and wireless modems. An example of OFDM has each of the K_b bits independently modulated on a separate subcarrier frequency, and the subcarrier frequencies are chosen to ensure the orthogonality. The format for an OFDM signal is

$$X_z(t) = \begin{cases} \displaystyle\sum_{l=1}^{K_b} X(l)\sqrt{\frac{E_b}{T_p}}\exp\left[j2\pi f_d(2l - K_b - 1)t\right] & 0 \le t \le T_p \\ 0 & \text{elsewhere} \end{cases} \quad (2.27)$$

with modulation symbols $X(l) = a(I(l))$, where $a(I(l))$ is the constellation mapping and $2f_d$ is the separation between adjacent subcarrier frequencies that are used to transmit the information. The transmission rate of this form of OFDM is $W_b = K_b/T_p$ bits per second. For clarity of discussion, the remainder of the section will assume all binary modulation mappings are Binary Phase Shift Keying (BPSK, i.e., $a(0) = 1$ and $a(1) = -1$). A more general form of OFDM could use any type of mappings and any number of bits per subcarrier. For example, 4 bits could be transmitted per subcarrier using a 16-ary modulation.

The OFDM transmitted waveform is a sum of K_b complex sinusoids. This transmitted waveform will have a complex envelope that changes significantly over the transmission time as the K_b complex sinusoids change in phase relative to each other. The larger the value of K_b, the larger this variation over the transmission time will be. An OFDM waveform has a significant difference between the peaks in amplitude and the average value of the amplitude. This high peak-to-average ratio requires the radios in an OFDM system to have a large dynamic range to process the signal without distortion.

The demodulator for OFDM is a matched filter for each subcarrier (due to the orthogonality between subcarriers). We will use $Y(k)$ to denote the matched filter output for the kth bit or subcarrier. The demodulator computes a filter output for each bit, $k \in \{1, \ldots, K_b\}$, and hence the complexity of the OFDM optimum demodulator is $O(K_b)$ as opposed to the $O(2^{K_b})$ for an arbitrary modulation that transmits K_b bits of information.

2.3.1.1 Spectral Characteristics of OFDM

To simplify the notation needed in this discussion we will make the following definition.

Definition 2.5 *The Fourier transform pair of the unit energy rectangular pulse function*

$$u_r(t) = \begin{cases} \dfrac{1}{\sqrt{T_p}} & 0 \le t \le T_p \\ 0 & \text{elsewhere} \end{cases} \tag{2.28}$$

is $U_r(f) = \sqrt{T_p}\,\text{sinc}\,(\pi f\, T_p) \exp\left[-j\pi f\, T_p\right]$.

Taking the Fourier transform of the OFDM signal (2.27) and using the frequency shift property of the Fourier transform gives

$$X_z(f) = \sum_{l=1}^{K_b} X(l) \sqrt{E_b}\, U_r\, (f - f_d(2l - K_b - 1)). \tag{2.29}$$

Further simplifications occur if each bit is equally likely. The spectrum in this case becomes

$$D_{X_z}(f) = \frac{E_b}{K_b} \sum_{l=1}^{K_b} |U_r\, (f - f_d(2l - K_b - 1))|^2. \tag{2.30}$$

This spectrum is plotted in Figure 2.4 for $K_b = 4$ and $f_d = 0.5/T_p$, where dotted lines refer to individual spectra and the solid line represents average spectrum, respectively. The conclusions we can draw about the average energy spectrum of OFDM is that the bandwidth occupancy for this type of an OFDM is proportional to $1/T_p$ and to K_b. Recall that the transmission rate is $W_b = K_b/T_p$ bits per second. Consequently, the transmission efficiency for binary OFDM is in the neighborhood of 1 bit/s/Hz, though the exact number will be a function of how the engineering bandwidth is defined.

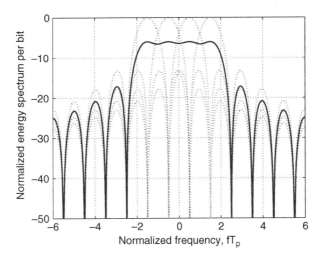

Figure 2.4 Average energy spectrum per bit for OFDM. $K_b = 4$, equally likely bits and BPSK modulation, with $f_d = 1/2T_p$.

2.3.1.2 Cyclic Prefix: Combating Frequency Selectivity

Most of the advantages of orthogonal modulations introduced so far come from the fact that orthogonality will be kept after passing the transmit signal through a frequency-flat channel, i.e., no intersymbol interference (ISI) will be introduced. This, however, will no longer be valid if the channel is frequency-selective[‡] and no preprocessing is imposed at the transmitter. Thus the MLWD has the complexity of $O(2^{K_b})$ due to the loss in orthogonality. Is it possible to maintain the demodulation complexity of $O(K_b)$ and yet provide good error performance over frequency-selective channels? The answer is yes, and there is a very simple modification of OFDM transmitted waveforms that permits simple suboptimum demodulators on frequency selective channels: adding a **cyclic prefix** to the OFDM signal. Mathematically we can show that

- The effect of frequency selectivity on an OFDM symbol is only to change the effective pulse shape for each bit.

- If we extend the time support of the OFDM symbol from $[0, T_d]$ to $[-T_h, T_d]$, where T_h is the time support of the channel impulse response, all the intercarrier interference (ICI) is removed from each

---

[‡] A more precise definition of frequency selectivity is provided in Section 2.5.1

subcarrier output and a simple bit-by-bit demodulator can be implemented. In practice, the cyclic prefix length is chosen to be bigger than the longest delay spread[§] experienced in operation.

It should be noted that adding a cyclic prefix and using the simple demodulator result in a loss in both spectral efficiency and fidelity but in many applications the gain in demodulator simplicity in a frequency-selective channel makes up for these losses. Consequently OFDM has found significant utility in wireless communication applications.

In conclusion, OFDM provides a method to implement multiple bit transmission with good fidelity, reasonable complexity, and spectral efficiency. The bit error probability performance of the OFDM scheme highlighted in this section gives the same bit error probability performance as BPSK used in isolation. The spectral efficiency of binary OFDM is roughly 1 bit/s/Hz. The complexity of the optimum receiver in a frequency-flat channel is $O(K_b)$. The same receiver can be used in a frequency-selective channel with little fidelity degradation if some simple modifications of the transmitted signal are imposed. Consequently, OFDM has found significant utility in engineering practice. Readers interested in a more detailed discussion on OFDM may refer to Heiskala and Terry.[28]

2.3.2 Orthogonal Code Division Multiplexing

A second commonly used modulation that admits a simple optimal bit demodulation is Orthogonal Code Division Multiplexing (OCDM). With OCDM each of the K_b bits is independently modulated with an orthogonal waveform. This orthogonal waveform is often termed the *spreading waveform*. OCDM is often used in wireless systems with multiple users, where each spreading waveform is associated with a different user. The typical format for an OCDM signal is

$$X_z(t) = \begin{cases} \sum_{l=1}^{K_b} X(l)\sqrt{E_b}s_l(t) & 0 \le t \le T_p, \\ 0 & \text{elsewhere} \end{cases} \tag{2.31}$$

[§] A more precise definition of delay spread is provided in Section 2.5.1.

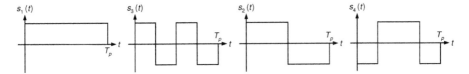

Figure 2.5   An example of spreading waveforms for OCDM and $K_b = 4$.

where $X(l) = a(I(l))$, and $s_l(t)$ is often denoted the spreading signal for the lth bit. Here we assume that both $E[|X(l)|^2] = 1$ and $E_{s_l} = 1$. The transmission rate of this form of OCDM is $W_b = K_b/T_p$ bits per second. It should be noted that OFDM is a special case of OCDM with $s_l(t) = \exp[j2\pi f_l t]/\sqrt{T_p}$.

There are a wide variety of ways to construct these spreading waveforms and one example is given in Figure 2.5 for $K_b = 4$. Certainly this construction is not unique. The time waveform of OCDM can have significant variations across the transmission time. An OCDM waveform can have a significant difference between the peaks in amplitude and the average value of the amplitude. This difference will increase as K_b increases. This high peak-to-average power ratio requires the radios in an OCDM system to have a high dynamic range to process the signal without distortion.

The optimal demodulator again has the form of K_b parallel single bit optimal demodulators. Restricting ourselves to BPSK modulation on each spreading waveform, the MLBD has the form

$$\Re\left[\int_0^{T_p} Y_z(t)s_k^*(t)dt\right] = \Re[Y(k)] \begin{array}{c} \hat{I}(k)=0 \\ > \\ < \\ \hat{I}(k)=1 \end{array} 0. \tag{2.32}$$

The demodulator computes a filter output for each bit, $k \in \{1, \dots, K_b\}$, and hence the complexity of the OCDM optimum demodulator is again $O(K_b)$ as opposed to the $O(2^{K_b})$ for an arbitrary modulation that transmits K_b bits of information.

2.3.2.1   Spectral Characteristics of OCDM

The spectral characteristics of OCDM provide additional insight. Using the same techniques as in OFDM and assuming for simplicity that each bit is equally likely and BPSK modulation is used, the spectrum in this case becomes

$$D_{X_z}(f) = \frac{E_b}{K_b} \sum_{l_1=1}^{K_b} |S_l(f)|^2, \tag{2.33}$$

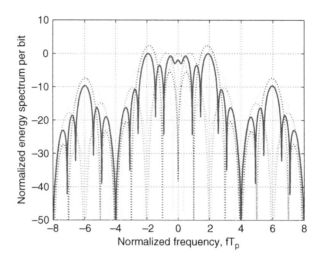

Figure 2.6 The average energy spectrum per bit for an OCDM spreading signal set defined in Figure 2.5. (See Color Plate 1)

where $S_l(f) = \mathcal{F}\{s_l(t)\}$. The important thing to note here is that the spectrum of OCDM waveforms is directly proportional to the spectrum of the chosen spreading waveforms. The spectrum of the set of spreading signals chosen in Figure 2.5 is shown in Figure 2.6 (again dotted lines for individual spectra and the solid line for average spectrum). The conclusions we can draw about the average energy spectrum of OCDM is that the bandwidth occupancy for this type of an OCDM is proportional to $1/T_p$ and to K_b. The bandwidth is greater than if a single bit was sent in isolation since K_b orthogonal waveforms need to be constructed for each of the K_b transmitted bits. Since in multi-user applications of OCDM the bandwidth of the spreading waveform is much larger than the transmission rate of the users, OCDM is often denoted *spread spectrum modulation*.[60]

2.3.2.2   Combating Frequency Selectivity
Similar to OFDM, frequency-selective channels will also cause the loss of orthogonality in an OCDM system. There are three different general methods to deal with this problem. One way is to still utilize the matched filter and operate with ISI caused by the frequency selectivity. In certain situations this interference can be tolerated. A second method is to try eliminating ISI by using some interference cancellation techniques.[60] The final way is to implement a multi-user demodulator at the receiver.[58] Interested readers can refer to these

books for a detailed treatment for OCDM systems operating in wireless channels.

2.3.3   Binary Stream Modulation

A third commonly used modulation that admits a simple optimal bit demodulation is orthogonal time division multiplexing. This is perhaps the most intuitive form of orthogonal modulation. Orthogonal time division multiplexing is used in a vast majority of digital communication systems in some form. The idea is simple: data are streamed in time (one bit following another). For the remainder of this chapter this orthogonal time division multiplexing will be referred to as *stream modulation*. With a stream modulation each of the K_b bits is independently modulated on the same carrier frequency with a time shifted waveform. The typical format for a stream modulation using linear modulation is

$$X_z(t) = \sum_{l=1}^{K_b} X(l)\sqrt{E_b}u(t - (l-1)T), \qquad (2.34)$$

where $X(l) = a\,(I(l))$, T is known as the symbol or bit time in stream modulations, and $u(t)$ is the unit energy pulse of length T_u. The transmission rate of stream modulation is $W_b = K_b/T_p$ bits per second. It should be noted that $T_p = (K_b - 1)T + T_u$ and for large K_b that $W_b \approx 1/T$.

This orthogonal time shift condition is often known as Nyquist's criterion for zero ISI.[35] The remaining question is how to design orthogonal time shifted waveforms. There is a wide variety of ways to construct these waveforms, but the simplest way is to limit $u(t)$ to only have support on $[0, T]$. For example, if $K_b = 4$, one can choose $T = T_p/4$ and have the set of time shifted waveforms as given in Figure 2.7. Certainly this construction is not unique, and most practical stream modulations use pulse shapes that are not time limited to T seconds as in Figure 2.7. It is interesting to note that with the waveforms chosen in Figure 2.7 the amplitude of the transmitted signal will be constant. The ability to more carefully control peak-to-average power ratio is one advantage of stream modulation.

The optimal demodulator again has the form of K_b parallel single bit optimal demodulators. Since the sample modulation format is repeated in time, the required filtering operation can also be serial with only one filter sampling several times. This demodulator is shown in Figure 2.8. The demodulator has one filter whose output is sampled for

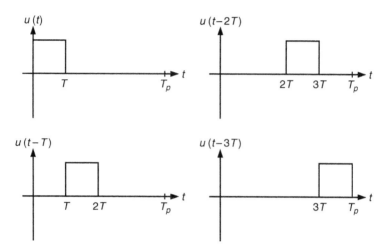

Figure 2.7 An example of time shifted pulses for a linear stream modulation and $K_b = 4$.

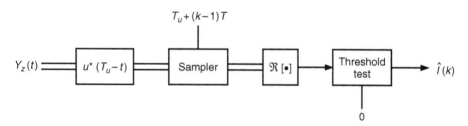

Figure 2.8 The optimal demodulator for stream modulations using BPSK.

each bit, $k \in \{1, \ldots, K_b\}$, and hence the complexity of the stream modulation optimum demodulator is again $O(K_b)$ as opposed to $O(2^{K_b})$. A characteristic of linear stream modulation that makes for an efficient implementation is that only one filter is needed to implement the demodulator. This pulse shape matched filter only needs to be sampled at different times to obtain sufficient statistics for demodulation.

The spectrum of stream modulations is only a function of the pulse shape. Similar techniques as used before in other orthogonal modulations give

$$D_{X_z}(f) = \frac{E_b}{K_b} \sum_{l_1=1}^{K_b} |U(f)|^2 = E_b G_u(f), \qquad (2.35)$$

where $U(f) = \mathcal{F}\{u(t)\}$ and $G_u(f) = |U(f)|^2$. The important thing to note here is that the spectrum of linear stream modulation is directly proportional to the spectrum of the chosen pulse shape. The spectrum of the linear stream modulation defined in Figure 2.7 is, by the use of Definition 2.4, $D_{X_z}(f) = E_b T \text{sinc}^2(fT)$. The conclusions we can draw about the average energy spectrum of stream modulation is that the bandwidth occupancy is also proportional to $1/T_p$ and to K_b.

In conclusion, orthogonal modulation is a technique where K_b bits are transmitted by using orthogonal waveforms for each bit. The orthogonality of the individual bit waveforms enables each bit to be detected optimally and serially with a complexity identical to the situation where the bit was transmitted in isolation. Examples of orthogonal modulation included in this chapter are modulations where orthogonality was obtained by frequency spacing (OFDM), complex waveforms (OCDM), or by time spacing (stream modulation). Orthogonal modulations address how to reduce the exponential complexity of optimum demodulation to linear complexity by an appropriate signal design. As a final note, most deployed communication systems use stream modulation in some form even when using OFDM or OCDM.

2.3.4   Orthogonal Modulations with Memory

The orthogonal modulation with memory (OMWM) is a general modulation that combines orthogonal modulation with a signal mapping having memory. The memory is normally included in the mapping to either improve the fidelity of demodulation or change the spectral characteristics of memoryless linear orthogonal modulation. OMWM incorporates most error control or spectrum control coding schemes.[8,32,61] Our goal in this chapter is not to explore how to design OMWM, but to overview the communication theory behind the performance, spectral efficiency, and demodulation complexity.

OMWM has modulation symbols generated by a finite state machine. The resulting modulation symbols can then be transmitted using a linear orthogonal modulation. The data modulation symbols, $\tilde{X}(l)$, are due to the stream of information bits, $I(l), l = 1, K_b$. The tilde notation will be used to differentiate between modulations that have memory (tilde) and memoryless modulations (not tilde). OMWM consists of a finite state machine operating at an integer fraction of the symbol rate, $1/N_m T$. The K_b bits to be transmitted are broken up into blocks of K_s in length (a total of $N_b = K_b/K_s$ blocks per frame). At each symbol time a new set of K_s bits, $\vec{I}(m)$, is input into a finite state machine, and this produces a new constellation label, $\vec{J}(m)$, and a new modulation state, $\sigma(m+1)$. The vector (length N_m) constellation label at the finite state machine output is used as an input to a modulator that produces an N_m symbol block of M_s-ary modulation symbols. The output modulation symbols are

$$\tilde{D}_z((m-1)N_m+i) = a(J_i(m)),\tag{2.36}$$

where $a(\cdot)$ is the constellation mapping, and $J_i(m)$ is the ith component of $\vec{J}(m)$. The output of the orthogonal modulation has the form

$$X_z(t) = \sum_l \tilde{D}_z(l)s_l(t).\tag{2.37}$$

To keep a consistent normalization, the mapping, $\tilde{D}_z(l) = a(J_i(m))$, is selected such that $E\left[\left|\tilde{D}_z(l)\right|^2\right] = R$. A total of N_b trellis transitions are needed to communicate the K_b bits. Figure 2.9 shows the block diagram for a general linear orthogonal modulation with memory. N_s again denotes the number of states in the modulation, and the non-linear equations governing the updates are

$$\sigma(m+1) = g_1\left(\sigma(m), \vec{I}(m)\right)\tag{2.38}$$

$$\vec{J}(m) = g_2\left(\sigma(m), \vec{I}(m)\right).\tag{2.39}$$

Note that $J(m) = 0, \ldots, M_s^{N_m} - 1$, where each component of $\vec{J}(m)$ only take values $J_i(m) = 0, \ldots, M_s - 1$, $i = 1, \ldots, N_m$. The constellation label at time m and in position i will generate the modulation symbol at time $(m-1)N_m + i$ for the orthogonal modulation. In general, it is

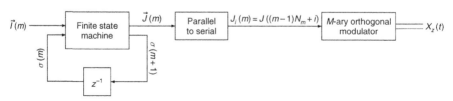

Figure 2.9 The block diagram for a general modulation with memory.

usually desirable to have ν_c extra symbols transmitted to return the modulation to a common final state at the end of the transmission frame. The total length of the frame for the orthogonal modulation with memory is denoted N_f; hence, $N_f = N_b N_m + \nu_c$, where ν_c is a code-dependent constant. Due to termination, the effective rate is $R_{eff} = \frac{K_b}{N_b N_m + \nu_c} = \frac{K_b K_m}{K_b N_m + \nu_c}$, but if a large number of bits are transmitted, then the rate becomes approximately $R = K_m / N_m$. If a rate less than one is desired, then a communication engineer would choose $N_m > K_m$. If a higher transmission rate is desired, then the communication engineer would choose $N_m < K_m$.

OMWM is used in practice to get better performance than orthogonal modulations or to change the spectral characteristics of orthogonal modulations while still maintaining a complexity that is $O(K_b)$. OMWM can be implemented in a wide variety of rates and complexities. For example, Figure 2.10 is a plot of the achieved spectral efficiency for the memoryless modulations and the operating points of several of the best OMWM that have appeared in the literature. The discussion of these modulations and the methods of demodulation requires a level of sophistication that is not appropriate for this chapter. Interested readers are referred to books on modern error control codes.[8,32,61] The important point is that Shannon's upper bound is one that is achievable in modern communications with a demodulation complexity that is $O(K_b)$ as desired.

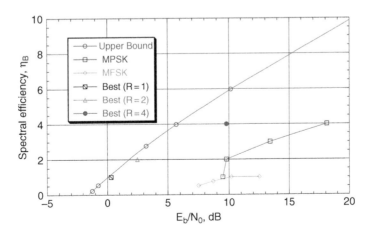

Figure 2.10 A comparison between the modulations considered in this chapter. (See Color Plate 2)

2.4 Propagation in Wireless Channels

A goal of this section is to provide a brief introduction to wireless channels. This is not meant to be a rigorous derivation based on electromagnetic theory. The goal here is much simpler: to introduce and motivate
models used in the development of wireless data communications. The
literature contains a myriad of terminology to reflect the complexities
of wireless propagation, e.g., Rayleigh fading, delay spread, frequency-
selective fading, time-varying fading, rich scattering environments,
and spatially independent fading. This section examines physical models and details these concepts. Throughout this chapter, the notation
is that the transmitted signal is $x_c(t)$ and the received signal is $y_c(t)$
in the passband and the corresponding signals in the baseband are
$x_z(t)$ and $y_z(t)$, respectively. Although the baseband representation of
signals is adequate to describe linear systems more simply in almost all
the cases, the passband representation is used in one example in this
section in order to explain the essence of time variation and its relation to mobility. Thus both notations are introduced in parallel. After
laying the necessary foundation, the baseband notation is exclusively
used throughout the rest of the discussion.

   This system engineering way of viewing the time-varying channel
is appropriate in that it models the waveforms that are input into
the antenna terminal at the transmitter and the waveforms that are
induced at the antenna output at the receiver. With this view of wireless propagation, we ignore the fact that electromagnetic vector fields
are induced in a three-dimensional space. Taking this simplistic view
does not enable us to discuss such topics as the polarization aspects
of antennas. While these issues are important, they are simply out of
scope for the current exposition.

   The wireless channel between any transmit antenna and receive
antenna can be modeled as a linear time-varying system. A simple
way of characterizing this linear time-varying system is to define its
time domain kernels, $h_c(t, \tau)$ and/or $h_z(t, \tau)$, in the passband and/or
the baseband, respectively. With the time domain kernel, the input–
output relationships are given as

$$Y_c(t) = R_c(t) + W_c(t) = \int_{-\infty}^{\infty} h_c(t, \tau) X_c(t - \tau) d\tau + W_c(t) \qquad (2.40\text{a})$$

$$Y_z(t) = R_z(t) + W_z(t) = \int_{-\infty}^{\infty} h_z(t, \tau) X_z(t - \tau) d\tau + W_z(t), \qquad (2.40\text{b})$$

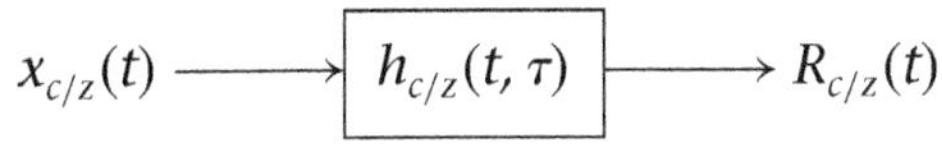

$$x_{c/z}(t) \longrightarrow \boxed{h_{c/z}(t, \tau)} \longrightarrow R_{c/z}(t)$$

Figure 2.11 The time–varying linear system representation of a wireless channel.

where $W_z(t)$ is a baseband additive noise process, and $W_c(t)$ is a bandpass additive noise process, the transformation between which is given as

$$W_c(t) = \Re \left\{ W_z(t) e^{j2\pi f_c t} \right\} \tag{2.41}$$

with f_c being the carrier frequency. This model is shown in Figure 2.11. The time variations in the wireless channel are captured with variable t, and the time dispersiveness is captured by the variable τ. The goal of this chapter is to briefly characterize this linear system and give some flavor for how propagation aspects can be characterized and used in communication systems. References that do a very good job of detailing out the systems perspective of wireless communications channels are 6, 10, 13, 18, 29, 37, 44, 50, 51 and 56.

The important idea in understanding wireless propagation is that a wireless transmission establishes a spatial standing wave in the environment. Once this idea is understood, the physics of the radio link can easily be interpreted to characterize wireless propagation. This section establishes the ideas of free space and multipath propagation. For simplicity, the discussion is limited to a two-dimensional propagation geometry but the generalization to three dimensions is straightforward yet tedious.[4,33,38,39] Angles in this two-dimensional geometry are defined relative to the positive x-axis.

2.4.1 Free Space Propagation

Free space propagation is characterized by a transmission where there are no obstacles between the transmitter and the receiver. The transmitter can be viewed as launching a wavefront that propagates in a spherical fashion outward from the transmitter. When there is no mobility, the wireless channel is a linear time-invariant system. It introduces a free space propagation loss, L_p, and a delay, τ_p. The free space propagation loss is due to the power of the transmission being spread across a bigger spherical wave; consequently, the loss L_p grows with distance. The time delay in radio wave propagation is due to the speed of light, c, and the distance between the transmitter and the receiver. If the transmitter and receiver are separated by a distance d, the time delay will be

$$\tau_p = \frac{d}{c}. \tag{2.42}$$

The passband kernel in this case becomes an impulse function of τ only

$$h_c(t, \tau) = L_p \delta(\tau - \tau_p) \tag{2.43}$$

and (2.40a) reduces to the convolution integral

$$R_c(t) = L_p \int_{-\infty}^{\infty} x_c(t - \tau) \delta(\tau - \tau_p) \, d\tau$$
$$= L_p x_c(t - \tau_p). \tag{2.44}$$

For example, if the transmitted signal is a tone $x_c(t) = \sqrt{2}\cos(2\pi f_c t)$, the received signal becomes $R_c(t) = L_p \sqrt{2}\cos(2\pi f_c(t - \tau_p))$. Defining the propagation phase shift to be

$$\phi_p = -2\pi f_c \tau_p = \frac{-2\pi f_c d}{c} = \frac{-2\pi d}{\lambda_c}, \tag{2.45}$$

where λ_c is the wavelength of the carrier, the received signal is given as

$$R_c(t) = L_p \sqrt{2}\cos(2\pi f_c t + \phi_p). \tag{2.46}$$

This propagation-induced phase shift is due to both the carrier frequency and the time delay from transmission to reception. It should be noted in most scenarios of interest (high carrier frequencies and large propagation distances) that the propagation phase shift is usually much larger than 2π so that later it will be well justified to model ϕ_p as a random phase uniform in $[0, 2\pi]$. We shall not apply a full mathematical derivation of the corresponding baseband kernel $h_z(t, \tau)$ given the results of (2.43). Instead, we use the following intuition. The complex envelope of a transmitted tone is $x_z(t) = 1$, and the received complex envelope (from (2.46)) is $R_z(t) = L_p \exp[j\phi_p] = H$. H is often referred to as a multiplicative distortion. Equivalently, if a modulated signal, $x_z(t)$, is transmitted via free space propagation, then the received signal has the form

$$R_z(t) = Hx_z(t - \tau_p), \tag{2.47}$$

which means that the baseband kernel in (2.40b) is

$$h_z(t, \tau) = H\delta(\tau - \tau_p) \tag{2.48}$$

and the integral in (2.40b) becomes also a convolution integral. Consequently, wireless propagation in free space with no mobility produces a multiplicative distortion and a time delay.

When there is mobility with free space propagation, the wireless channel between the transmit and receive antennas potentially becomes a linear time-varying system. This time variation is basically due to the change of the distance between the transmitter and the receiver with time. This is reflected, in turn, to a change in the transmission delay and the path loss with time as well. However, in practice, the variation in path loss can be neglected as it is very slight during the transmission time span. The distance between the transmitter and the receiver, as a function of time, is denoted by $d(t)$, and the corresponding transmission delay is denoted by $\tau_p(t)$. If a signal is radiated from the transmit antenna at time t, it will impinge on the receive antenna at time $t + \tau_p(t)$ (i.e., when the distance between the transmitter and the receiver becomes $d(t + \tau_p(t))$), where τ_p is now a time-dependent transmission delay given by

$$\tau_p(t) = \frac{d\left(t + \tau_p(t)\right)}{c}. \tag{2.49}$$

For simplicity, we assume that the variation of $d(t)$ with time is due only to the motion of the receiver away from the stationary transmitter as shown in Figure 2.12(a).[*] The passband kernel of the channel in this case is expressed as

$$h_c(t, \tau) = L_p\delta\left(\tau - \tau_p(t)\right). \tag{2.50}$$

Substituting $h_c(t, \tau)$ in (2.40a), we get

$$R_c(t) = L_p x_c\left(t - \tau_p(t)\right). \tag{2.51}$$

In practical applications, $v(t)$ can be approximated to be uniform (i.e., $d(t) = d + vt$, where d is the distance between the transmitter and the

---

[*] But transmitter mobility can be taken into consideration using the concept of relative velocity. If the transmitter moves with a velocity $\vec{v}_t(t)$ and the receiver with a velocity $\vec{v}_r(t)$, then this situation is equivalent to a stationary transmitter and a receiver moving with a velocity $\vec{v} = \vec{v}_r(t) - \vec{v}_t(t)$.

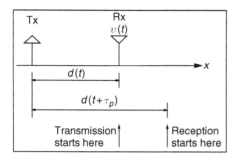

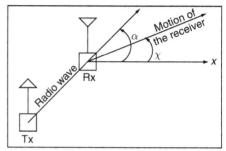

(a) Receiver moving away from a fixed
transmitter with a constant velocity v.

(b) Receiver moving in a scenario with
arbitrary angle of arrival α and direction
of motion given by χ.

Figure 2.12 A model for a time-varying channel with mobility from the receive side only.

receiver at time $t = 0$) if its variation during the transmission time is not significant. From (2.49), this gives a transmission delay

$$\tau_p(t) = \frac{d + vt + v\tau_p(t)}{c} \qquad \left(\text{which implies } \tau_p(t) = \frac{d + vt}{c - v}\right), \qquad (2.52)$$

and a received signal

$$R_c(t) = L_p x_c\left(t - \frac{d}{c - v} - \frac{vt}{c - v}\right). \qquad (2.53)$$

Because $c \gg v$, the approximation ($\frac{1}{1-x} \approx 1 + x$ when $x \ll 1$) is usually used to simplify $R_c(t)$ as

$$R_c(t) \approx L_p x_c\left(t - \frac{d}{c}\left(1 + \frac{v}{c}\right) - \frac{vt}{c}\right). \qquad (2.54)$$

By comparing this result with (2.44), we see that they are similar in having a time delay term. But due to the relative motion with respect to the wave propagation, a new term $\frac{v}{c}$ arises in (2.54), which causes an effect called the **Doppler frequency shift**, or simply the **Doppler shift**. The name arises from the fact that the transmission of an unmodulated carrier $x_c(t) = \sqrt{2}\cos(2\pi f_c t)$ results in a received signal $R_c(t) = L_p\sqrt{2}\cos\left(2\pi(f_c - f_D)t + \tilde{\phi}_p(f_D)\right)$, where

$$f_D = f_c\frac{v}{c}, \quad \text{and} \quad \tilde{\phi}_p(f_D) = -2\pi(f_c + f_D)\tau_p, \qquad (2.55)$$

i.e., the received carrier is frequency shifted from the transmitted one by f_D. For a reason to appear shortly, f_D is called the **maximum Doppler shift**. If we now apply the same approach to a two-dimensional case where the direction of motion of the receiver is allowed to differ from the direction of the arriving wavefront as shown in Figure 2.12(b), it is straightforward to see that

$$R_c(t) = L_p \sqrt{2} \cos \left(2\pi \left(f_c + f_d \right) t + \tilde{\phi}_p \left(-f_d \right) \right), \qquad (2.56)$$

where the Doppler shift becomes

$$f_d = -f_D \cos \left(\alpha - \chi \right) \qquad (2.57)$$

and α and χ are respectively the angles subtended by the direction of arrival of the radio wave and the direction of motion of the receiver with the positive x-axis. It is now clear why f_D is called the maximum Doppler shift; it is the maximum value that f_d can attain. Obviously, Figure 2.12(a) is a special case of the model in Figure 2.12(b) with $\alpha = \chi = 0$. The received complex envelope is

$$R_z(t) = \tilde{H} \exp \left[j2\pi f_d t \right]. \qquad (2.58)$$

$\tilde{H}$ has the same magnitude as H; $\left| \tilde{H} \right| = |H| = L_p$, but $\arg \left\{ \tilde{H} \right\} = \tilde{\phi}_p \left(-f_d \right)$. This makes the baseband kernel in (2.40b) equal to

$$h_z(t, \tau) = \tilde{H} \exp \left[j2\pi f_d t \right] \delta(\tau - \tau_p). \qquad (2.59)$$

This system is not linear time invariant since a frequency of f_c is put into the system and a frequency of $f_c + f_d$ comes out of the system. This frequency difference is due to the motion of the antenna through the time-varying standing wave established by the radio transmission. As it moves through space the receiving antenna will output a waveform that varies at a different frequency due to the Doppler effect. It is important to realize that the rapidity of time variation is directly proportional to the speed of motion and the carrier frequency. Higher frequency operation and/or higher speed mobility will make the channel vary faster. The rapidity of the time variation is also a function of the difference between the angle of the incoming radio wavefront and the direction of motion. Motion directly into the wavefront produces the highest Doppler shift. Motion away from the wavefront produces the smallest Doppler shift. Motion perpendicular to the wavefront produces no Doppler shift. This is the first case where the geometry of the propagation produces differences in the received signal characteristics. This geometric dependence of wireless channels is frequently exploited in communication practice.

Similarly, when a modulated carrier is transmitted, the received complex envelope can be well approximated as

$$R_z(t) = \tilde{H} \exp\left[j2\pi f_d t\right] x_z(t - \tau_p). \tag{2.60}$$

Therefore, wireless propagation in free space with mobility produces a time-varying multiplicative distortion and a time delay. The time variations in the multiplicative distortion are entirely due to the Doppler shift.

The time-varying nature of wireless channels is due to the motion of the link environment. In the simple case of free space propagation, either transmit antenna motion or receive antenna motion can cause this time variation. The idea that a tone transmission from an antenna establishes a standing wave in space is a very important concept in understanding wireless channels and will be used to illustrate many important concepts in the sequel.

2.4.2 Multipath Propagation

Wireless channels are interesting from a theoretical point of view due to the inherent multipath propagation characteristics. Wireless communication occurs in physical environments, and these environments typically have objects that diffract or scatter radio waves. Since the radio wavefronts are diffracted and scattered by objects in the environment, there will be multiple ways that electromagnetic energy can propagate from the transmitter to the receiver. This multipath propagation is what makes radio communications interesting. The multipath modes of propagation will have a received signal much like the free space channels described previously.

First, let us consider a multipath propagation environment with no mobility. Electromagnetic waves will add at each position in space so the resultant voltage observed on a receiver antenna will be the superposition of the effects of each multipath signal. The model of this multipath propagation is

$$R_z(t) = \sum_{n=1}^{m_p} H_n x_z(t - \tau_n), \tag{2.61}$$

where m_p is the number of distinct multipaths in a particular channel, H_n is the multipath multiplicative distortion, and τ_n is the multipath time delay. Note that each multipath will have a complex multiplicative distortion. It should be apparent that wireless channels can be very different in terms of the number of multipaths, the amplitude distributions, and the time delays that are produced. For example, channels seen in indoor wireless local area computer networks have significantly different characteristics than channels seen in outdoor mobile telephone networks. The remainder of this chapter explores the impact of these differences on how digital communications is accomplished on such channels. Again, when the channel is linear and time invariant, it has kernel

$$h_z(t, \tau) = h_z(\tau) = \sum_{n=1}^{m_p} H_n \delta(\tau - \tau_n). \tag{2.62}$$

Multipath channels without mobility are well modeled as linear time-invariant systems. There are some very important characteristics in this multipath channel that help in understanding wireless communications. The amplitude of the multiplicative terms, H_n, depends on, for example, the distance between the transmitter and the receiver, the types of media that the radio waves propagate through, and the surfaces of the scattering objects. The phase is a function of all these same things and also the multipath propagation length from the transmitter to the receiver, i.e., $\arg\{H_n\} = \theta_n - 2\pi f_c \tau_n = \theta_n - \frac{2\pi d_n}{\lambda_c}$, where d_n is the distance traveled from the transmit antenna to the receive antenna by the nth multipath signal, and θ_n is the residual phase. Consequently, the phase of a multipath multiplicative distortion can change, significantly when the path length changes a small distance (note: small in this discussion is in terms of wavelengths of the carrier frequency). The path delays, τ_n, are only a function of the path length as each delay is only a function of how long the radio wave took to travel from the transmitter to the receiver. The exploitable characteristic in wireless channels that has led to a rapid growth in techniques to communicate over wireless channels is that the phase changes rapidly with path length change while the time delays and amplitudes change slowly with path length change. The sequel will explore how these characteristics are exploited in wireless communications.

It is instructive to consider the case where a tone is transmitted. Recall that a tone transmission is modeled with $x_z(t) = 1$. The received signal in this case will be

$$R_z(t) = \sum_{n=1}^{m_p} H_n. \qquad (2.63)$$

Hence the received signal is also a tone with an amplitude and phase determined by the phasor sum of the multipath multiplicative distortions. A pictorial representation of this phasor sum is given in Figure 2.13 for $m_p = 4$. Wireless communication is challenging because even in situations where each of the multipath amplitudes are large, there is a possibility that the phasor sum might be close to zero. This situation, where the resultant amplitude is small, even when the amplitudes of each of the individual paths are relatively large, is often denoted as a **fade**. A significant number of the techniques used in wireless digital communications are implemented to mitigate the effects of fading. The general idea is to transmit over redundant channels to provide **diversity** to reduce the impact of fades.

A multipath propagation environment with mobility has a similar extension as the multipath environment with no mobility. The only difference between the mobility and the non-mobility case is that each multipath will have a Doppler shift. The Doppler shift from (2.57) is a function of both the multipath arrival angle, α_n, and the direction of motion, χ. In direct analogy to the single path case, we denote

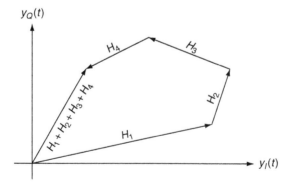

Figure 2.13 The phasor diagram of the multipath response to a tone transmission. $m_p = 4$.

the Doppler frequency for the nth path as $f_n = -f_D \cos(\chi - \alpha_n)$. The multipath model with mobility for the received signal with a tone transmission is given as

$$R_z(t) = \sum_{n=1}^{m_p} \tilde{H}_n \exp[j2\pi f_n t]. \qquad (2.64)$$

In the case where the receiver antenna is the one in motion, it should be noted that when multipaths arrive from different angles, different Doppler frequencies can be produced. Here, the channel is again linear but time-varying and has a kernel given as

$$h_z(t, \tau) = \sum_{n=1}^{m_p} \tilde{H}_n \exp[j2\pi f_n t] \delta(\tau - \tau_n). \qquad (2.65)$$

This simple channel model can accurately represent wireless channels with mobility.

2.4.3   Gaussian Modeling of Multipath Channels

Wireless channels are often modeled by system designers as random quantities. This randomness is meant to capture the great variability in a channel physical geometry for each possible radio deployment. Random quantities are often characterized by statistical averages. Since most wireless channels are characterized by complex quantities, the discussion here will assume complex-valued random variables. The most commonly used statistical quantities for characterizing random channels are the mean, $m_x = E[X]$; the average power, $P_x = E[|X|^2]$; the variance, $\text{var}(X) = 2\sigma_X^2 = E[|X - m_x|^2]$; and the correlation coefficient. The correlation coefficient between two complex random variables, X and Y, is defined as

$$\rho_{XY} = \frac{E[(X - m_x)(Y - m_y)^*]}{2\sigma_X \sigma_Y}. \qquad (2.66)$$

Since wireless communication techniques try to find redundant and independent channel realizations, ensuring a small correlation coefficient between channel realizations is often a goal.

Models for the random channel realization have been a topic of continued interest since the inception of wireless communications. For simplicity, the case of a single transmitted tone is considered. Recall that the received signal in this case has the form

$$R_z(t) = \sum_{n=1}^{m_p} H_n = H_z. \tag{2.67}$$

To evaluate the performance, it is often necessary to characterize the resultant channel gain, H_z. Since $H_z = H_I + jH_Q$ is due to a sum of m_p random variables and m_p is large, the central limit theorem becomes applicable.[31] Recall that the central limit theorem implies that if all of the path gains do not vary greatly in distribution then the resultant complex channel gain will converge to a complex Gaussian random variable as m_p gets large. Furthermore, since for most wireless channels the phase of each multipath gain is equally likely to be anywhere on $[-\pi, \pi]$, this implies that $E[H_z] = 0$. Note that if $H_z = H_I + jH_Q$ is a zero mean complex Gaussian random variable, $H_A = |H_z| = \sqrt{H_I^2 + H_Q^2}$ is a Rayleigh random variable. **Rayleigh fading** refers to the case where H_z is modeled as a zero mean complex Gaussian random variable. The probability density function (PDF) of a Rayleigh-distributed random variable with variance 0.5 is plotted in Figure 2.14.

Rayleigh fading is an often-used model in wireless communications. First, the Gaussian model is conducive to getting results in analysis. The assumption of a Gaussian model for the fading statistics has

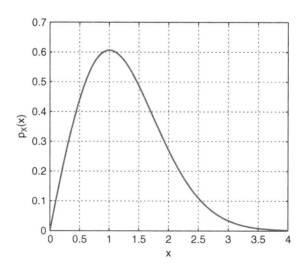

Figure 2.14 Theoretical PDF of a real-valued Rayleigh-distributed random variable of variance 0.5.

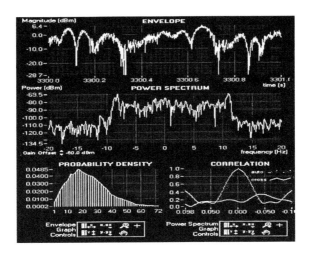

Figure 2.15   A field test in downtown Columbus, OH.[11]

been the starting point for a near innumerable number of theoretical papers. Second, the zero mean Gaussian model represents a worst-case scenario. If no paths dominate the sum, then there is typically not a line-of-sight path from the transmit antenna to the receive antenna. If there is a large number of paths, the transmitter and receiver are embedded in a dense scattering environment. The situation of dense scattering and no line-of-sight is usually the worst-case scenario for wireless communications. The tradition in digital communications is to design the system to meet the performance requirements in the worst-case scenario knowing that better channels will produce better performance. Finally, the Rayleigh model is accurate in many situations. For example, Figure 2.15 shows the results of a field test in downtown Columbus, OH.[11] The plot in the bottom left-hand corner is a histogram of the channel amplitudes measured during this particular trial. This histogram looks very much like the Rayleigh PDF in Figure 2.14. This combination of being analytically simple, a worst-case model, and an accurate model for measured data makes the Rayleigh model very popular in wireless communications.

2.5   The Selectivity of Wireless Channels

This section gives a brief overview of wireless channel selectivity and the terminology used to describe it. The behavior of a single radio channel, one transmit antenna to one receive antenna, is explored.

The literature often refers to this as Single-Input Single-Output (SISO) communications. This very focussed presentation is only intended to enable the reader to get a quick understanding of the important channel characteristics in modern wireless systems. More details from a systems perspective can be found in references 6, 10, 13, 18, 29, 37, 50, 51 and 56.

This section shows that

1. Wireless channels are frequency-selective (the channel response will vary with the frequency of the transmitted signal).

2. Wireless channels are spatially selective (the channel response will vary with the position of the transmitter and/or the receiver antenna).

3. Wireless channels where there is motion in the channel are time-selective (the channel response will vary with time).

All three selectivities play a significant role in modern wireless communication.

2.5.1 Frequency Selectivity

Wireless channels are frequency-selective (the channel response will vary with the frequency of the transmitted signal). Recall in the case of no mobility that the channel kernel for this linear time-invariant system is

$$h_z(t, \tau) = h_z(\tau) = \sum_{n=1}^{m_p} H_n \delta(\tau - \tau_n), \tag{2.68}$$

where the time delays are arranged such that $\tau_1 < \tau_2 < \cdots < \tau_{m_p}$. The transfer function of the channel is given as

$$H_z(f) = \mathcal{F}\{h_z(\tau)\} = \sum_{n=1}^{m_p} H_n \exp[-j2\pi f \tau_n]. \tag{2.69}$$

A typical impulse response (magnitude) and a magnitude squared transfer function of a wireless channel are shown in Figure 2.16. The time delays and corresponding gains of the various multipaths are shown in Table 2.1.

It is clear from (2.69) that the transfer function for a channel with no mobility is a sum of sinusoids. This sum of sinusoids produces a

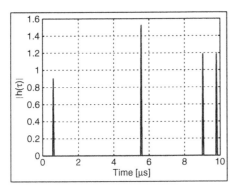
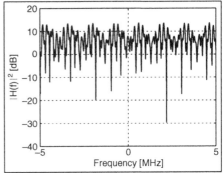

(a) Impulse response magnitude   (b) Transfer function

Figure 2.16 Impulse response and transfer function of a typical wireless channel with $m_p = 4$. Delays and gains of the various multipaths are given in
Table 2.1.

Table 2.1 Time delays and gains of the
four multipaths of the channel in
Figure 2.16.

τ_1	$0.64\mu s$	H_1	$-0.5106 + j0.7435$
τ_2	$5.60\mu s$	H_2	$-0.5217 + j1.4320$
τ_3	$9.08\mu s$	H_3	$-0.8701 + j0.8075$
τ_4	$9.84\mu s$	H_4	$0.8479 - j0.8299$

transfer function that varies with frequency. A parameter often used
to characterize the amount of frequency selectivity is **delay spread**.
Equation (2.16) is rewritten after taking the complex exponential with
the smallest delay, τ_1, as a common factor:

$$H_z(f) = \exp\left[-j2\pi f\tau_1\right]\left(H_1 + \sum_{n=2}^{m_p} H_n \exp\left[-j2\pi f(\tau_n - \tau_1)\right]\right). \qquad (2.70)$$

it is clear that the term that varies fastest with frequency in the transfer
function sum of sinusoids is due to an argument of $\tau_d = \tau_{m_p} - \tau_1$. τ_d
is defined to be the **delay spread**, and the delay spread completely
controls how quickly the channel transfer function amplitude changes
with respect to frequency. For example, in the channel considered
in Figure 2.16, the delay spread is about $9.2\mu s$ and an oscillation at
about a $100\,$kHz rate is apparent in the transfer function. When a

communication engineer considers how to design systems for wireless communication, delay spread is one of the first channel parameters to be identified.

In many situations, it is important to understand how the channel response changes as a function of frequency. The typical measure of how the channel changes with frequency is the frequency domain correlation function

$$R_H(\Delta_f) = E\left[H_z(f)H_z(f - \Delta_f)\right]. \tag{2.71}$$

Often, data communication utilizes different frequencies to communicate redundant information. When doing this it is important to identify the separation that is needed to get the channels to behave somewhat independently. Engineers use the concept of **coherence bandwidth** to characterize this frequency diversity. Coherence bandwidth is often defined to be the frequency separation when the frequency domain correlation function drops below a certain level.

The frequency domain correlation function can be simplified when considering the models discussed in this chapter. Since each multipath is due to a separate propagation path, it is a good model to assume each path is independent of the other paths. Since the phase is due partially to path delay, and typical path delays cause a phase rotation that is many times greater than 2π, it is a good approximation to assume that $E[H_n] = 0$. After plugging (2.69) into (2.71) and using these two approximations, one can show that

$$R_H(\Delta_f) = E\left[\sum_{n=1}^{m_p} |H_n|^2 \exp\left[-j2\pi\Delta_f \tau_n\right]\right]. \tag{2.72}$$

Hence, the correlation function is a function of the distribution of the number of paths, the distribution of the path delays, and the distribution of the path powers. If the number of paths and the random delays in the channel are independent of the multipath gains, then we have

$$R_H(\Delta_f) = E_{m_p, \tau_n}\left[\sum_{n=1}^{m_p} P_H(\tau_n) \exp\left[-j2\pi\Delta_f \tau_n\right]\right], \tag{2.73}$$

where $P_H(\tau)$ is the average power of the path with a delay τ. This expectation and summation are often[42] approximated** as

$$R_H(\Delta_f) = \int_0^\infty S_H(\tau) \exp\left[-j2\pi\Delta_f\tau\right] d\tau, \qquad (2.74)$$

where $S_H(\tau)$ is denoted in the literature as the **power delay profile**. One can think about $S_H(\tau)$ as a power density which is a function of delay. Hence, there is a nice duality with the standard notion of a correlation function in time and a power spectral density in frequency. The important point to remember is that the power distribution in delay will determine how the correlation of the channel transfer function will change with frequency. Channels that have a larger delay spread will vary more rapidly in the frequency domain. For instance, an outdoor land mobile radio (mobile phones) typically will have a frequency response that will change more rapidly with frequency than an indoor wireless local area network. The coherence bandwidth for an outdoor radio is around 100 kHz and the coherence bandwidth for an indoor radio is around 4 MHz.[44]

In certain situations, the effect of the channel on the transmitted signal is essentially frequency-flat. When the product of the bandwidth of the signal and the delay spread is much less than unity, the channel can be modeled in this way. If this is the case, all multipaths arrive at roughly the same time, τ_p, compared to the time variations of the transmitted signal. Consequently, the received signal on each antenna can be accurately modeled as

$$R_z(t) = \sum_{n=1}^{m_p} H_n x_z(t - \tau_n) = x_z(t - \tau_p) \sum_{n=1}^{m_p} H_n = H_F x_z(t - \tau_p). \qquad (2.75)$$

In other words, for a channel modeled as frequency-flat, the received signal from each antenna can be modeled as the transmitted signal multiplied by a complex fading distortion.

Examples with different signal bandwidths clarify the concepts of frequency flatness and frequency selectivity. A time domain example where the delay spread is much larger than the reciprocal of the bandwidth is shown in Figure 2.17. The transmitted signal is a pulse which has a spectral square root raised cosine shape,[42] and the channel in Figure 2.16 is used. Figure 2.17-b shows a plot of the I and Q signals at the channel output. It is clear from the output in Figure 2.17(b)

** In some special cases the approximation can be shown to be exact.

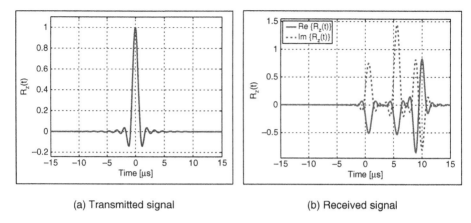

(a) Transmitted signal   (b) Received signal

Figure 2.17 A time domain example of a communication system where the channel can be modeled as frequency-selective. The channel is given in Figure 2.16.

that the different paths can be resolved and the received signal must be modeled with $\sum_{n=1}^{m_p} H_n x_z(t - \tau_n)$. On the other hand, a time domain example where the delay spread is much smaller than the reciprocal of the bandwidth is shown in Figure 2.18. Again, the channel is taken to be the one in Figure 2.16. When looking at the output in Figure 2.18(b), it is clear that the different paths are not resolvable and the received signal can be modeled with $\sum_{n=1}^{m_p} H_n x_z(t - \tau_p) = H_F x_z(t - \tau_p)$. Note that Figures 2.17(b) and 2.18(b) show the output to have in-phase

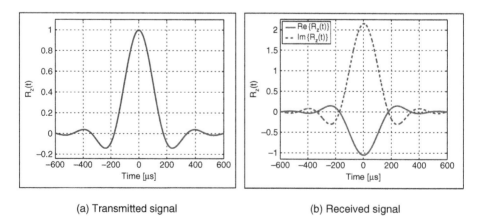

(a) Transmitted signal   (b) Received signal

Figure 2.18 A time domain example of a communication system where the channel can be modeled as frequency-flat. The channel is given in Figure 2.16.

and quadrature components although the input is real, because the multiplicative distortion, H_F, is complex-valued. Frequency-flat fading is a good model in narrowband communications and is also useful in OFDM where each subcarrier pulse is usually designed to see a frequency-flat channel.

2.5.2   Spatial Selectivity

Wireless channels are spatially selective (the channel response will vary with the position of the transmitter and/or the receiver antenna). The discussion will concentrate on the case where the transmit antenna stays stationary and the receive antenna is the one that shifts positions.

The important idea in understanding how standing waves are established in space is to realize that the received multipath signals are all potentially coming from different directions. Figure 2.19 shows a pictorial representation of this situation. For simplicity, assume that the signal bandwidth is small enough such that the channel is well modeled as frequency-flat. Define a reference position, A_1, at which the received signal has the form

$$R_z(t, A_1) = x_z(t - \tau_p) \sum_{n=1}^{m_p} H_n(A_1) = H_F(A_1) x_z(t - \tau_p). \qquad (2.76)$$

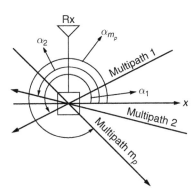

Figure 2.19   A graphical representation of the angle of arrival of received multipath waveforms.

Recall that the propagation phase shift for each multipath is

$$\phi_n = \theta_n - 2\pi f_c \tau_n = \theta_n - \frac{2\pi d_n}{\lambda_c}, \tag{2.77}$$

where θ_n is a phase that is a function of the propagation path, and d_n is the distance traveled by the nth multipath signal. This form shows that the phase for each multipath can change significantly as the receive antenna position varies in space (changing d_n). If when repositioning the receive antenna the phases of all the multipaths change the same amount, the signal amplitude will not change. Whereas, if the phases of the multipaths change a different amount, the resulting amplitude can be quite different. An example of a phasor diagram to represent the multipath summation is given in Figure 2.21. In Figures 2.21(a) and 2.21(b), the amplitudes of the corresponding multipaths are equal. The only difference between the two diagrams is the path gain phases. Since the phases change in a different fashion for each path gain, the resultant amplitude at the two locations is quite different. This phasor diagram demonstrates how a spatial standing wave can be established.

Here, we consider the mathematical details of moving the receive antenna to a different location A_2. The geometry of the problem considered is shown in Figure 2.20. The antenna spacing is denoted with Δ, and the angle between the line connecting A_1 and A_2 and the positive x-axis is β. To simplify the analysis, the receive antenna is assumed to be in the far field so that the incoming rays are parallel (which is quite accurate when the distance between the scatterer and the receive

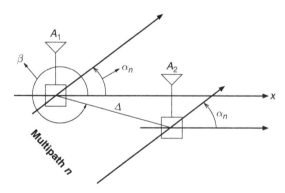

Figure 2.20 The geometry of two spatially separated receive antennas for multipath n.

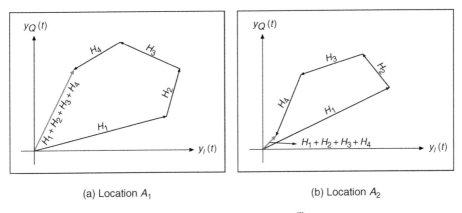

(a) Location A_1 (b) Location A_2

Figure 2.21   A vector diagram to represent the $\sum_{n=1}^{m_p} H_n(A_i)$ at two positions.

antenna is greater than $\frac{2\Delta^2}{\lambda}$).[†] With this formulation, the nth path will have a path length that is different from the path to the reference antenna by an amount

$$\Delta_n = -\Delta \cos(\beta - \alpha_n).\tag{2.78}$$

Note that if the antenna is moved toward the arriving ray, the path length becomes shorter. If the antenna is moved away from the arriving ray, the path length becomes longer. If the antenna is moved perpendicular to the arriving ray, the path length does not change. This change in path length will result in a change in phase of each multipath signal

$$\arg\{H_n(A_2)\} - \arg\{H_n(A_1)\} = \frac{2\pi\Delta\cos(\beta - \alpha_n)}{\lambda_c}.\tag{2.79}$$

As a result, the received signal at position A_2 has the form

$$R_z(t, A_2) = x_z(t - \tau_p)\sum_{n=1}^{m_p} H_n(A_1)\exp\left[\frac{j2\pi\Delta\cos(\beta - \alpha_n)}{\lambda_c}\right].\tag{2.80}$$

It is seen from (2.80) that the variability in space is due entirely to how $\frac{2\pi\cos(\beta - \alpha_n)}{\lambda_c}$ varies across the received multipath signals.

If the wavefronts seen at the receiver are coming from a wide variety of directions, the spatial standing wave will vary significantly over

---

[†] This model concentrates on the effect of the last scatterer only. Previous paths connecting other scatterers remain the same.

a wavelength spatial scale. For instance, if the paths are assumed to arrive uniformly in angle on the interval $[-\pi, \pi]$, the received signal at position A_2 will be given as

$$R_z(t, A_2) = x_z(t - \tau_p) \sum_{n=1}^{m_p} H_n(A_1) \exp\left[\frac{j2\pi\Delta \cos\left(\beta - \frac{2\pi(n-1)}{m_p}\right)}{\lambda_c}\right]. \qquad (2.81)$$

To gain insight into how the channel varies with spatial separation, we can look at the spatial correlation function

$$R_H(\Delta) = E\left[H_F(A_1)H_F^*(A_1 - \Delta)\right]. \qquad (2.82)$$

Assuming that the multipath gains are independent and zero mean, the correlation function reduces to

$$R_H(\Delta) = E_{m_p}\left[\sum_{n=1}^{m_p} E\left[|H_n(A_1)|^2\right] \exp\left[\frac{j2\pi\Delta \cos\left(\beta - \frac{2\pi(n-1)}{m_p}\right)}{\lambda_c}\right]\right]. \qquad (2.83)$$

If the average power of the multipath gains are constant as a function of angle of arrival, then

$$R_H(\Delta) = E_{m_p}\left[\frac{2\sigma_H^2}{m_p} \sum_{n=1}^{m_p} \exp\left[\frac{j2\pi\Delta \cos\left(\beta - \frac{2\pi(n-1)}{m_p}\right)}{\lambda_c}\right]\right]. \qquad (2.84)$$

If m_p is large, this correlation function will converge to [29]

$$R_H(\Delta) = \frac{2\sigma_H^2}{2\pi} \int_{-\pi}^{\pi} \exp\left[\frac{j2\pi\Delta \cos(\theta)}{\lambda_c}\right] d\theta = 2\sigma_H^2 J_0\left(\frac{2\pi\Delta}{\lambda_c}\right), \qquad (2.85)$$

where $J_0(x)$ is the Bessel function of the first kind order zero.[1] It should be apparent from (2.85) that the correlation coefficient between the channel gains of two antennas spaced by Δ is $\rho_H(\Delta) = J_0\left(2\pi\frac{\Delta}{\lambda_c}\right)$. From this characterization, it is clear that when the multipaths arrive from a wide variety of angles, the signal at the antenna output will change significantly after moving the antenna as little as a wavelength of the radio carrier frequency. For example, Figure 2.22 is a plot of a sample function of the standing wave in space in an environment where the multipaths are coming from a wide variety of directions. The behavior of this sample function corroborates the analytical result given in (2.85). If the wavefronts seen at the receiver are coming from roughly

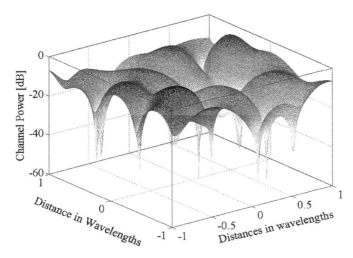

Figure 2.22   A sample function of the amplitude of a standing wave in space with isotropic angles of arrival.

the same direction, the spatial standing wave will vary more slowly in space. Let us consider as a degenerate case the situation when all the paths come roughly from the same direction, α. The channel gain at the second antenna location has the form

$$
H_T\,(A_2) = \sum_{n=1}^{m_p} H_n(A_1)\exp\left[\frac{j2\pi\cos(\beta-\alpha)}{\lambda_c}\right]
$$

$$
= \exp\left[\frac{j2\pi\cos(\beta-\alpha)}{\lambda_c}\right]\sum_{n=1}^{m_p} H_n(A_1)
$$

$$
= H_T(A_1)\exp\left[\frac{j2\pi\cos(\beta-\alpha)}{\lambda_c}\right]. \tag{2.86}
$$

It is clear that the change in the resulting multiplicative distortion is only a phase shift, but the standing wave in space does not have a varying amplitude.

Two quantities are often used to characterize propagation environments: angle spread, α_s, and the mean angle of arrival, α_m. For this discussion the **angle spread** of a propagation environment is given as

$$
\alpha_s = \max_{n_1,n_2=1,\ldots,m_p}\left(\alpha_{n_1}-\alpha_{n_2}\right). \tag{2.87}
$$

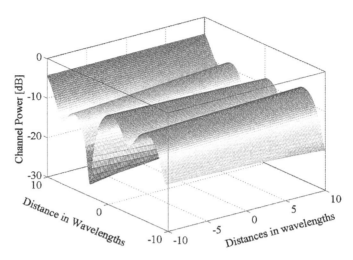

Figure 2.23 A sample function of the amplitude of a standing wave in space with $\alpha_m = 0°$ and $\alpha_s = 10°$.

In situations where there is an attempt to statistically model propagation, the angle spread has a more statistical meaning (i.e., standard deviation of the random arrival angle). Similarly, if the angle spread of the arriving waveforms is fairly small, the received signal will vary slowly in space. Both the angle spread and the orientation of the antenna displacement with respect to the mean angle of arrival will impact the characteristics of the spatial standing wave. Figure 2.23 shows a sample spatial standing wave for a mean angle of arrival equal to 0° and an angle spread of 10°. In examining this standing wave, one should note that the spatial field varies more in one direction (along the y-axis) than in the perpendicular direction (along the x-axis). This difference is because the 10° angle spread makes a greater variation in the received phases when the arriving angle is broadside to the antenna displacement. This characteristic results because the displacement perpendicular to α_m, $\frac{2\pi \sin(\alpha_n)}{\lambda_c}$, produces a greater range of phases for the received angle spread and, consequently, a greater amplitude variation than the displacement parallel to α_m, $\frac{2\pi \cos(\alpha_n)}{\lambda}$.

When the angle spread is small, many factors impact the characteristics of the spatial standing wave.

1. The wavelength will determine the amount of phase change that is produced with the spatial offset.

2. The angle spread determines how much variation there is in the phase of each of the multipath signals.

3. The angle difference between the mean angle of arrival and the antenna displacement determines the size of the relative variation in the phase of each of the multipath signals.

These characteristics are important to understand how antenna arrays should be constructed in multiple antenna modems.

The geometric dependence of the spatial standing wave leads antenna arrays to be designed differently depending on the propagation geometry that is likely to be seen in the wireless system. For instance, in cellular systems (macro cells) the base station towers are typically very high above the urban landscape. The typical propagation geometry that is seen in this situation is a significant amount of local scattering near the mobile (at street level) and not much other scattering, especially near the base station antennas. This has led base station antenna arrays to be designed such that antenna spacings are on the order of $10\lambda_c$. Alternatively, at the mobile unit the scattering environment is rich and often isotropically distributed. Because of this, antennas in array for mobile radios usually have less than a wavelength separation. Similarly, wireless local area network (WLAN) deployments often have antenna arrays where the antenna separation is on the order of wavelengths. In indoor environments typical of WLAN deployments, the scattering is much more often isotropically distributed and not so often concentrated with a small angle spread. The geometric scattering environment in which a multiple antenna wireless system is deployed must be considered carefully when designing the antenna array.

2.5.3   Time Selectivity

Wireless channels where there is motion in the channel are time-selective (the channel response will vary with time). The important thing to remember is that wireless channels are multipath channels and that motion produces a Doppler shift in a received multipath signal. Recall that the received channel gain for a transmitted tone $(x_z(t) = 1)$ in a time-varying wireless channel with mobility has the form

$$R_z(t) = \sum_{n=1}^{m_p} H_n \exp\left[j2\pi f_n t\right]. \qquad (2.88)$$

This implies that the received signal on a single antenna when the antenna is in motion is a sum of sinusoids. Recall that the frequencies of the sinusoids are a function of the angle of arrival compared to the angle of motion and the maximum Doppler frequency, f_D. The sum of sinusoid characteristic is frequently seen in measured field data. For example, examining Figure 2.15 from the field test in Columbus, OH, it is clear that the amplitude of the received signal (the top plot) is well modeled by a sum of sinusoids.

In many situations, it is important to understand how the channel response changes as a function of time. The typical measure of how the channel changes with time is the time domain correlation function defined as

$$R_H(\Delta t) = E[H_z(t)H_z(t - \Delta t)]. \tag{2.89}$$

Often, data communications utilize different time instances to communicate redundant information. When doing this, it is important to identify the time separation that is needed to get the channel gains to behave somewhat independently. Engineers use the concept of **coherence time** to characterize this time diversity. Coherence time is often defined to be time separation when the time domain correlation function drops below a certain level.

Since the time domain correlation function is a familiar tool for the communication engineer,[31] further insight is obtained using the traditional tools. The Fourier transform of the time domain correlation function is often denoted as the **Doppler spectrum**, i.e.,

$$S_H(\nu) = \mathcal{F}\{R_H(\Delta t)\}. \tag{2.90}$$

This Doppler spectrum provides the traditional frequency domain interpretation of the random time variation in the channel gains.

Different distributions on the angle of arrival will produce different types of time-varying fading. For instance, if the angle of arrival of each of the multipaths is isotropically distributed, a sample path of an example channel gain is shown in Figure 2.24(a) and the measured power spectral density of this sample path is shown in Figure 2.24(b). In this isotropic propagation environment, it is clear that the channel response will change fairly rapidly and the spectrum can have power over the full range of $[-f_D, f_D]$. **Doppler spread** is defined as $f_s = f_{max} - f_{min}$. It is clear from Figure 2.24(a) that in the case of isotropic fading, $f_s = 2f_D$. Alternatively, consider the case where the angle spread on the received signal is limited to 10°. A sample path of an example

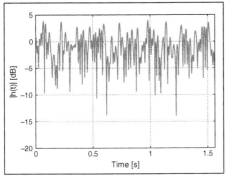

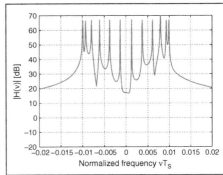

(a) Amplitude of the sample path

(b) Power spectrum of the sample path versus normalized frequency (T_S is the symbol duration)

Figure 2.24 A sample path of a fading process for an isotropic scattering environment. $f_D T_s = 0.01$.

channel gain for this limited angle spread is shown in Figure 2.25(a), and the measured power spectral density of this sample path is shown in Figure 2.25(b) for $\alpha_m = -30°$ and $\beta = 60°$. Here, in spite of having the same f_D, the channel gain in the narrow angle spread case clearly varies less rapidly than the channel gain for the isotropic scattering environment. This slow variation is due entirely to the narrow angle spread producing a smaller f_s. To characterize the behavior of

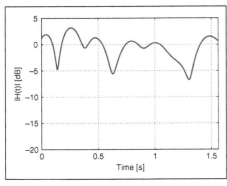

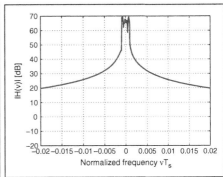

(a) Amplitude of the sample path

(b) Power spectrum of the sample path versus normalized frequency (Ts is the symbol duration)

Figure 2.25 A sample path of a fading process for a narrow angle spread. $\alpha_m = -30°$, $\beta - \alpha_m = 90°$, $\alpha_s = 10°$, and $f_D T_s = 0.01$.

the random channel gain over time, it is clear that both the vehicle speed and the angle spread of the propagation environment should be characterized.

It is worth examining the detailed form of the time domain correlation function. As in the two previous sections, we will assume that each of the path gains are independent and zero mean so that the time correlation reduces to

$$R_H(\Delta t) = E_{m_p \alpha_n} \left[\sum_{n=1}^{m_p} E\left[|H_n|^2\right] \exp\left[j2\pi f_n \Delta t\right] \right]. \tag{2.91}$$

Again, if the power of the paths are the same then this correlation function reduces to

$$R_H(\Delta t) = 2\sigma_H^2 E_{m_p \alpha_n} \left[\sum_{n=1}^{m_p} \exp\left[j2\pi f_D \cos\left(\beta - \alpha_n\right) \Delta t\right] \right]. \tag{2.92}$$

Note the analogy with the form for the spatial correlation. Consequently, when the scattering is isotropic and the number of paths gets large, Clarke showed that the correlation function will take the form[16]

$$R_H(\Delta t) = J_0\left(2\pi f_D \Delta t\right) \tag{2.93}$$

which can also be derived from (2.85) when motion with uniform velocity is assumed, i.e., $\Delta = v \times \Delta t$. This correlation function of Clarke is a very good model for an average channel but the instantaneous channel can be different. The general form for a correlation function generally behaves quite like this average; for example, the correlation function that was measured in Figure 2.15 (lower right figure) looks quite like (2.93). The important point to realize from the analytical result of Clarke is that the time variability of a fading process is entirely determined by the Doppler spread, and the Doppler spread is also entirely determined by the angle spread and the motion velocity.

2.5.4   Summary of Channel Characteristics

Having examined the characteristics of wireless channels leads to a final model that captures all of the important characteristics of wireless channels. The received signal is given as

$$R_z(t, A) = \sum_{n=1}^{m_p} H_n \exp\left[j2\pi f_D \cos\left(\beta - \alpha_n\right) t + j2\pi \frac{\Delta}{\lambda_c} \cos(\beta - \alpha_n) \right] x_z(t - \tau_n),$$

$$(2.94)$$

where A represents the antenna position, H_n represents the multipath gains at a reference position in space, Δ is the distance A is from the reference position, β is the angle of displacement of A from the reference position, α_n is the multipath arrival angle, and τ_n is the path delay. This model is very accurate, but is valid only for receiver antenna motion and receiver antenna displacement. It is very important to realize that this scenario is very limited in terms of all the possible channel realizations one would encounter in practice. For example, in wireless communications, often the transmitter may move, both transmitter and receiver may move, or even the environment may be moving. While generalizations of the above model for these other important cases are straightforward (if not tedious), they will not be pursued in this chapter as the important characteristics of the wireless channel will not change significantly in these other situations.

In summary, wireless channels are:

1. Frequency-selective
2. Spatially selective
3. Time-selective.

The details of each selectivity are a function of the following channel characteristics:

1. Power–delay profile
2. Mean angles of arrival/departure
3. Angle spread of arrival/departure
4. Motion of the transmitter/receiver/environment
5. Antenna array geometry.

System designers must be aware of how the channel geometry and the signal design will impact the performance of wireless communications.

2.6   Physical Models of Wireless Systems

This section provides the notation for the information-theoretic analysis of wireless systems in the sequel. Performance analysis of communication systems, in general, requires a mathematical model to describe the channel behavior and the signal and noise input-output relation. This modeling takes into account time, frequency, and spatial selectivities of the channel presented in Section 2.5, the later when multiple antennas are used. Several scenarios are discussed in this section to cover SISO and multiple antenna radio. A few common assumptions to all these scenarios are:

1. A vector of K_b bits $\vec{I} = [I(1) \; I(2) \; \cdots \; I(K_b)]^T$ is transmitted in T_p seconds.

2. The transmitted bit vector is first processed by an encoder to produce the transmit symbol vector $\vec{X}$ of N_f symbols $\vec{X} = [X(1) \; X(2) \; \cdots \; X(N_f)]^T$ as shown in Figure 2.26. The symbols $\{X(l)\}$ are selected from a (complex) M-ary alphabet and normalized such that $E\left[|X(l)|^2\right] = 1 \; \forall l \in \{1, \ldots, N_f\}$.

3. The number of transmit and receive antennas is denoted by L_t and L_r, respectively. The time signal output from transmit antenna n is denoted by $X_z^{(n)}(t)$, and that output from receive antenna m is denoted by $Y_z^{(m)}(t)$.

4. Orthogonal modulations from Section 2.3 are used on all L_t transmit antennas to transmit $\vec{X}$. Transmit antenna n uses a vector of orthogonal waveforms $\vec{s}_n(t) = \left[s_{n,1}(t) \; s_{n,2}(t) \; \cdots \; s_{n,N_f}(t)\right]^T$ to transmit $\vec{X}$ in the following fashion:

$$X_z^{(n)}(t) = \sum_{l=1}^{N_f} X(l) \sqrt{E_s} s_{n,l}(t) = \sqrt{E_s} \vec{s}_n(t)^T \vec{X}, \quad 0 \le t \le T_p. \quad (2.95)$$

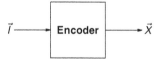

Figure 2.26   Mapping of bits to symbols.

It is usually desirable to make the notation more compact by grouping all the transmitted signals in one vector $\vec{X}_z(t) = \left[X_z^{(1)}(t) \right.$ $\left. X_z^{(2)}(t) \cdots X_z^{(L_t)}(t) \right]^T$ and all the waveform vectors in one $L_t \times N_f$ matrix $\mathbf{s}(t) = \left[\vec{s}_1(t) \ \vec{s}_2(t) \cdots \vec{s}_{L_t}(t) \right]^T$, the relation between which is

$$\vec{X}_z(t) = \sqrt{E_s}\mathbf{s}(t)\vec{X}. \tag{2.96}$$

5. The received signal at antenna m, $Y_z^{(m)}(t)$, is the superposition of all the transmitted signals after being processed by the channel in the fashion shown in (2.40b). If we denote the channel kernel between transmit antenna n and receive antenna m as $h_z^{(m,n)}(t,\tau)$, $Y_z^{(m)}(t)$ is expressed as

$$Y_z^{(m)}(t) = \sum_{n=1}^{L_t} \int_{-\infty}^{\infty} h_z^{(m,n)}(t,\tau) X_z^{(n)}(t-\tau)\,d\tau + W_z^{(m)}(t) \tag{2.97}$$

$$= \int_{-\infty}^{\infty} \vec{h}_m(t,\tau)^T \vec{X}_z(t-\tau)\,d\tau + W_z^{(m)}(t) = R_z^{(m)}(t) + W_z^{(m)}(t), \tag{2.98}$$

where $\vec{h}_m(t,\tau) = \left[h_z^{(m,1)}(t,\tau) \ h_z^{(m,2)}(t,\tau) \cdots h_z^{(m,L_t)}(t,\tau) \right]^T$. Grouping all the received signals in one vector following the same procedure as before yields

$$\vec{Y}_z(t) = \int_{-\infty}^{\infty} \mathbf{H}_z(t,\tau)\vec{X}_z(t-\tau)\,d\tau + \vec{W}_z(t) \tag{2.99}$$

with $\vec{Y}_z(t) = \left[Y_z^{(1)}(t) \ Y_z^{(2)}(t) \cdots Y_z^{(L_r)}(t) \right]^T$ and $\mathbf{H}_z(t,\tau) = \left[\vec{h}_1(t,\tau) \right.$ $\left. \vec{h}_2(t,\tau) \cdots \vec{h}_{L_r}(t,\tau) \right]^T$.[‡]

It is worth mentioning that for a multiple antenna system with fixed L_t, L_r, K_b, T_p and N_f, the performance of the system with respect to capacity and bit error performance depends on the design of the encoding technique $\vec{I} \to \vec{X}$ and the matrix $\mathbf{s}(t)$. In literature, the design

---

[‡] The superscripts (n) and (m) will be ignored when a SISO channel is considered as it is a special case with $L_t = L_r = 1$. For spatial selectivity on the receive side only (i.e., $L_t = 1$), a single superscript (m) is used.

of $\mathbf{s}(t)$ is termed space–time block coding (STBC).[§] A lot of work has been done to assess the performance and design of $\mathbf{s}(t)$ and the encoding technique. The reader may refer to references 2, 5, 23, 24, 26, 46, 52 and 53 for a few examples of design methodologies.

Enough notation is now available to express physical models of wireless transmission. Considering a general case, all three kinds of selectivity might exist. However, in many situations, the rate of channel variations in time is very low in comparison to the signaling rate which validates treating the channel as time-flat rather than time-selective. Similarly, wireless channels can be treated as frequency-flat if the rate of variation in frequency is much lower than the signaling bandwidth. The dependency of $\mathbf{H}_z(t, \tau)$ on t and τ captures this selectivity according to Table 2.2. In the following, models for different cases are presented. Each kind of channel is denoted by its flatness/selectivity in space, frequency, and time. SISO channels, however, are denoted by the last two only as spatial selectivity does not exist. For SISO channels, all vectors and matrices reduce to scalars.

2.6.1 Time-Flat Frequency-Flat (TF/FF) Channels

The simplest SISO channel is one where variation in time and frequency is negligible. This channel is a good model to a low mobility, low bandwidth (compared to the coherence bandwidth) digital communication. From Table 2.2, the received signal becomes

Table 2.2 Time and frequency selectivities in the structure of $\mathbf{H}_z(t, \tau)$.

Time-flat, frequency-flat	$\mathbf{H}_z(t, \tau) = \mathbf{H}_z \delta(\tau)$	$\vec{Y}_z(t) = \mathbf{H}_z \vec{X}_z(t) + \vec{W}_z(t)$
Time-selective, frequency-flat	$\mathbf{H}_z(t, \tau) = \mathbf{H}_z(t) \delta(\tau)$	$\vec{Y}_z(t) = \mathbf{H}_z(t) \vec{X}_z(t) + \vec{W}_z(t)$
Time-flat, frequency-selective	$\mathbf{H}_z(t, \tau) = \mathbf{h}(\tau)$	$\vec{Y}_z(t) = \int_{-\infty}^{\infty} \mathbf{h}(\tau) \vec{X}_z(t - \tau) d\tau + \vec{W}_z(t)$

[§] The "time" dimension comes from the fact that stream modulation is used. In the same sense, if OFDM is used, we can have space-frequency block coding.

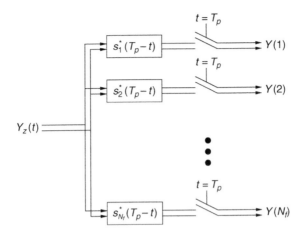

Figure 2.27   The optimal demodulator for orthogonal modulation.

$$Y_z(t) = H_z X_z(t) + W_z(t), \tag{2.100}$$

where H_z is a scalar, and $W_z(t)$ is a white complex Gaussian noise process with zero mean and variance N_0. The optimal demodulator consists of N_f matched filters as illustrated in Figure 2.27. The output of the kth matched filter sampled at $t = T_p$ is

$$Y(k) = \int_0^{T_p} Y_z(\lambda) s_k^*(\lambda) \, d\lambda, \qquad k = 1, \ldots, N_f \tag{2.101a}$$

$$= \sum_{l=1}^{N_f} H_z X(l) \sqrt{E_s} \int_0^{T_p} s_l(\lambda) s_k^*(\lambda) \, d\lambda + N(k) \tag{2.101b}$$

$$= \sqrt{E_s} H_z X(k) + N(k), \tag{2.101c}$$

where (2.101c) comes from the orthogonality of $\{s_l(t)\}$. The discrete time noise $N(k)$ is related to $W_z(t)$ by

$$N(k) = \int_0^{T_p} W_z(\lambda) s_k^*(\lambda) \, d\lambda. \tag{2.102}$$

So, it has a zero mean, it is white, and its variance is equal to N_0 due to the unit norm of $\{s_l(t)\}$. The matched filter outputs $\{Y(k)\}$ represent a set of sufficient statistics about the transmitted symbols $\{X(k)\}$. It is worth mentioning here that for a TF/FF channel, all orthogonal modulation techniques lead to the same results regarding the output $Y(k)$. In the sequel, a TF/FF channel is modeled according to (2.101c).

2.6.2 Time-Varying Frequency-Flat (TV/FF) Channels

The TV/FF channel models narrowband communication with mobility. Due to time variations, the channel fading profile cannot be assumed flat throughout the transmission time T_p. Stream modulation can be used in this situation to conquer the time selectivity as follows. The waveforms of stream modulation have the form

$$s_k(t) = u(t - (k-1)T), \tag{2.103}$$

where the pulse $u(t)$ spans an interval T_u such that $T_p = T_u + (N_f - 1)T$. If T is small enough compared to the coherence time of the channel, then the variation of the channel within an interval of T will be insignificant. From Table 2.2,

$$Y_z(t) = H_z(t)X_z(t) + W_z(t). \tag{2.104}$$

As explained in Section 2.3.3, a single matched filter can represent the optimal demodulator by sampling its output at multiples of T. The sampled outputs are expressed as

$$Y(k) = \int_0^{T_p} Y_z(\lambda) u^*(\lambda - (k-1)T) d\lambda, \qquad k = 1, \ldots, N_f \tag{2.105a}$$

$$= \sum_{l=1}^{N_f} \sqrt{E_s} X(l) \int_0^{T_p} H_z(t) u(\lambda - (l-1)T) u^*(\lambda - (k-1)T) d\lambda + N(k). \tag{2.105b}$$

If $H_z(t)$ were constant, the Nyquist criterion of the pulse shape would eliminate all ISI, i.e.,

$$\int_0^{T_p} u(\lambda - (l-1)T) u^*(\lambda - (k-1)T) d\lambda = \delta_{k-l}. \tag{2.106}$$

As a first level approximation, the impact of the time-varying channel on the ISI will be ignored and the integral can be modeled as

$$\int_0^{T_p} H_z(\lambda) u(\lambda - (l-1)T) u^*(\lambda - (k-1)T) d\lambda = H_z[k]\delta_{k-l}, \tag{2.107}$$

where $H_z[k] = H_z(kT)$. This simplifies the outputs of the matched filter as

$$Y(k) = \sqrt{E_s} H_z[k] X(k) + N(k), \tag{2.108}$$

where $N(k)$ has the same characteristics as in the case of a TF/FF channel.

2.6.3 Time-Flat Frequency-Varying (TF/FV) Channels

The case of a TF/FV channel models wideband communication with no mobility. Frequency selectivity can be avoided similar to time selectivity by using an OFDM transmission. Following a similar analysis, it is easily shown that

$$Y(k) = \sqrt{E_s}\mathcal{H}_z[k]X(k) + N(k), \qquad (2.109)$$

where

$$\mathcal{H}_z[k] = \mathcal{F}\{H_z(t)\}|_{f=\frac{k}{T_0}} = \int H_z(t)e^{-j2\pi\frac{k}{T_0}t}\,dt \qquad (2.110)$$

is the value of the transfer function of the channel at the kth subcarrier frequency, and T_0 is the OFDM symbol time. It should be noted that the mathematical models for a TV/FF channel with stream modulation and the TF/FV channel with OFDM are identical. Only the physics that cause the variations in $\mathcal{H}_z[k]$ and $H_z[k]$ are different.

2.6.4 Receiver-Space-Varying Frequency-Flat (RSV/FF) Channels

Receiver spatial selectivity is exploited in systems that use multiple antennas at the receive side. Therefore, this model holds for systems employing receiver diversity only. This can be easily extended to the frequency-varying case similar to earlier sections. Figure 2.28 shows

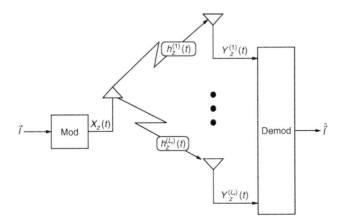

Figure 2.28 A communication system with receiver diversity using L_r antennas.

such a communication system. Each of the receive antennas is connected to a chain of N_f matched filters whose outputs are sampled
at $t = T_p$ as in Figure 2.27. Ideally, the placement of the receive
antennas should allow as little correlation* between the channel gains
$\vec{H}_z[k] = \left[H_z^{(1)}[k], \ldots, H_z^{(L_r)}[k] \right]^T$ as possible to make use of space diversity. Referring to Figures 2.22 and 2.23, it is obvious that the narrower
the spread of the angle of arrival, the larger the required spacing
is between the antennas. However, in practice, a correlation coefficient between the channel gains of about 0.6 is enough to yield good
performance.[19,47] Using the results of Table 2.2 and following similar steps as in Sections 2.6.1, and 2.6.2, the received symbol vector is
expressed as

$$\vec{Y}(k) = \sqrt{E_s} \vec{H}_z^T X(k) + \vec{N}(k) \tag{2.111}$$

for an RSV/TF/FF channel and

$$\vec{Y}(k) = \sqrt{E_s} \vec{H}_z^T[k] X(k) + \vec{N}(k) \tag{2.112}$$

for an RSV/TV/FF channel**, where $\vec{Y}(k) = [Y_1(k) \cdots Y_{L_r}(k)]^T$, $\vec{N}(k) = [N_1(k) \cdots N_{L_r}(k)]^T$ and $N_m(k)$, are independent identically distributed
(IID) Gaussian random variables with zero mean and variance N_0.

2.6.5 Transmitter-Receiver Space-Varying Frequency-Flat (TRSV/FF) Channels

The most general case of space diversity is to have both transmitter
and receiver employing multiple antennas. Such a model is extensively
used in space-time coding systems. Figure 2.29 depicts a TRSV/TF/FF
channel. From Table 2.2, it is straightforward to show that

$$\vec{Y}(k) = \mathbf{H}_z \vec{X}(k) + \vec{N}(k) \tag{2.113}$$

for a TRSV/TF/FF channel and

$$\vec{Y}(k) = \mathbf{H}_z[k] \vec{X}(k) + \vec{N}(k) \tag{2.114}$$

---

* This is equivalent to saying that $E\left[\vec{H}_z[k] \vec{H}_z^{\dagger}[k] \right] = \mathbf{I}_{L_r}$ is desired.
** In this case, only one matched filter per receiver antenna is required whose outputs are
sampled at multiples of T.

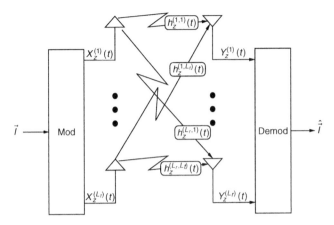

Figure 2.29 A communication system with L_t transmit antennas and L_r receive antennas.

for a TRSV/TV/FF channel. The models presented for SISO and multiple antenna systems, although simple, represent a powerful tool in understanding the channel behavior. For example, the dimensions of $\mathbf{H}_z[k]$ corresponding to L_r and L_t highly influence the channel capacity. The behavior of the entries $H_z^{(m,n)}[k]$ may admit or prohibit certain equalization and detection techniques. Section 2.7 treats wireless systems from an information-theoretic point of view. Capacities of various channels are computed based on the models introduced above.

2.6.6   Paradigms for Wireless Communication

It is clear from (2.113) and (2.114) that for reliable detection the receiver and/or the transmitter need to have estimates of the channel state information (CSI) $\mathbf{H}_z[k]$. Otherwise, correct amplitude and phase of the transmitted symbols can never be restored. The paradigm used in wireless communication is highly dependent on the availability and reliability of the transmitter/receiver CSI estimates (TCSI/RCSI). When $\mathbf{H}_z[k]$ is unknown or partially known, it is modeled as a random variable.[9,22,49] The random variable corresponding to TCSI is denoted by $\hat{\mathbf{H}}_T$, and that corresponding to RCSI is denoted by $\hat{\mathbf{H}}_R$. A general model for wireless communication that highlights the aspects of CSI on the modulation and demodulation is shown in Figure 2.30. TCSI and/or RCSI are assumed to be perfect if $\hat{\mathbf{H}}_T$ (respectively, $\hat{\mathbf{H}}_R$) equals $\mathbf{H}_z$. No TCSI/RCSI is available if $\hat{\mathbf{H}}_T$ (respectively, $\hat{\mathbf{H}}_R$) is independent

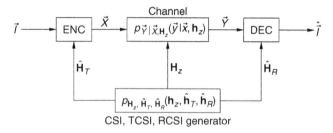

Figure 2.30   Block diagram of a communication system with TCSI/RCSI.

of H_z. The communication paradigm that is implemented in practice is heavily influenced by what level of CSI is available. For example, in wireless local area computer networks, it is assumed that mobility is low and that channel information can be made available at the transmitter. High mobility systems (e.g., mobile phones) typically do not try to feedback precise CSI. Table 2.3 lists some of the common assumptions about channel knowledge at the transmitter and receiver in practice. The most common assumption in practice is when RCSI is perfect and the transmitter only knows the statistics of the channel. For further information about how the communicating parties estimate the CSI, the reader can refer to references 14, 24, 34 and 63.

Table 2.3   Estimation of the channel state at the transmitter and the receiver.

Transmitter	CSI	Instantaneous and perfect: $\hat{H}_T[k] = H_z[k]$
		Instantaneous and noisy: $\hat{H}_T[k] = H_z[k] + n[k]$
		Delayed: $\hat{H}_T[k] = H_z[k-q]$ $q = 1, 2, \ldots$
	Only statistics of CSI	$p_{\hat{H}_T}\left(\hat{h}_T\right) = p_{H_z}(h_z)$, and independent
Receiver	CSI	Instantaneous and perfect: $\hat{H}_R[k] = H_z[k]$
		Instantaneous and noisy: $\hat{H}_R[k] = H_z[k] + n[k]$
		Delayed: $\hat{H}_R[k] = H_z[k-q]$ $q = 1, 2, \ldots$
	Only statistics of CSI	$p_{\hat{H}_R}\left(\hat{h}_R\right) = p_{H_z}(h_z)$, and independent

2.7 Modern Wireless Communication

The characteristics of wireless channels introduced in Section 2.5 have greatly influenced the development of wireless communications. There has been a clear evolution of thinking in how wireless communication systems are designed, and the goal for the remainder of this chapter is to motivate this evolution with the understanding derived from the physical models of wireless propagation. The evolution of wireless can be summarized with the following "ages":

1. **Wireless communications is unreliable**: In this age, engineers attempted a direct application of standard communication theory to wireless multipath environments. Wireless multipath communications produced channels that were frequency, time, and spatially selective. Some positions in frequency, time, and space supported reliable communication and some did not if the channel experienced a deep fade. This unreliability was frustrating to users of digital communications, but it was viewed as a characteristic of wireless multipath communications.

2. **Redundancy can improve reliability**: In this age, engineers came to see that, if a channel was faded at a particular frequency, time, and/or position, more reliable communication could be achieved if redundant information was sent on another channel in frequency, time, and space. This redundancy in the dimensions of frequency, time, and space was termed diversity in wireless communications. For example, the Global System for Mobile communication (GSM) came to dominate mobile phone deployments because it used time interleaving, frequency hopping, wideband transmission, and multiple antennas to greatly improve the reliability of mobile phone communications.

3. **Spatial redundancy does not take bandwidth**: Pioneering work by Telatar[54] and Foschini and Gans[20] showed that radio systems with multiple antennas on both sides of the link can actually both increase the reliability of communication and simultaneously increase the spectral efficiency. This remarkable observation has spurred a major revolution in wireless communications. One good example of the fruits of this revolution is the IEEE 802.11n standard which promises spectral efficiencies over 10 bits/s/Hz.

The goal in the remainder of the chapter is to support the engineering justification for why these ages evolved. The justification is from an information-theoretical perspective.

A brief overview of diversity techniques will clarify some of the important issues. Section 2.5 shows that, in a multipath propagation environment, even when average signal power is large, certain points in frequency, time, and/or space will experience a deep signal fade. Diversity techniques provide independent replicas of the transmitted signal. This redundancy is intended such that when one or more such replicas is rendered useless due to deep fading, the others can compensate for it and enable reliable communications. Examples of diversity techniques include:

Frequency diversity: In frequency-selective channels, the same signal is retransmitted on different frequency subbands. The separation between these subbands must be at least equal to the channel coherence bandwidth so that, if the channel impulse response (CIR) on some of these subbands is significantly low, there is still enough redundancy to convey the transmitted signal to the receiver. Wideband transmission using OFDM is one example that uses frequency diversity.

Time diversity: In time-selective channels, the transmission of the symbols is repeated in different time slots separated by at least the channel coherence time. Consequently, the channel gains in these time slots are independent, so they convey independent versions of the transmitted signal.

Spatial diversity: Spatial diversity includes **transmit diversity** and **receive diversity** and exploits the spatial selectivity of wireless channels. In transmit diversity, more than one transmit antenna contribute to transmitting the signal. The receive antenna receives the sum of all these replicas weighted by the corresponding channel gains. In space-time coding systems, a cooperation scheme between the transmit antennas is deliberately designed to offer diversity gain. In receive diversity, on the other hand, the transmitted signal is redundantly received by more than one receive antenna.

Polarization diversity: Antennas with different polarizations participate in transmitting or receiving wireless signals to achieve diversity. Due to the different polarizations, they can have independent impulse responses even when their locations are very close. It should be noted that due to space constraints the introduction in this chapter does not really discuss propagation in a level of detail where polarization is understood.

This qualitative introduction gives a preliminary flavor of diversity and multiple antenna systems. Sections 2.5 and 2.6 give a sense of how these fades can be mitigated in frequency, time, and space. For instance, the concept of spatial selectivity introduced in Section 2.5.2 is the reason for exploiting space diversity. If the channel were spatially flat (i.e., not selective), it would be of no use employing more than one receive antenna as all of them would see identical channels. The sequel develops the theory of wireless communications roughly in chronological order of the ages highlighted above.

2.7.1   Capacity of the SISO Channel

Historically the first wireless data systems used on multipath channels were traditional SISO systems, and hence characterization of wireless systems in this chapter will start here. The early systems often supported only small data rates as early computer processor speeds could not utilize high data rate transfers. Therefore, the frequency-flat and time-flat channel model often was valid for these early systems. The capacity of the TF/FF SISO channel will also serve as a baseline for future comparison with multiple-antenna channel capacity. At the same time, several important concepts in fading channel capacity analysis such as *average* and *outage capacities* will be introduced. It should be noted that as a brief introduction most results will be given without proof. Readers can refer to an excellent tutorial on single-antenna fading channel capacity.[9] Using the results in Section 2.6, the received signal can be represented as

$$Y(k) = \sqrt{E_s} H_z X(k) + N(k), \qquad (2.115)$$

where H_z is the complex channel gain. As a first insight into wireless communication, this section will only consider the case where the receiver has perfect knowledge of H_z and the transmitter only knows the distribution of H_z.

Fixing transmission at a fixed location in frequency, time, and space allows the use of traditional Shannon theory. Traditional Shannon capacity only deals with a fixed channel gain $H_z = h$, and the traditional Shannon capacity is defined as the maximum mutual information between the channel input and output. This is proved to be equal to the maximum data rate that can be transmitted reliably over this channel. Thus, the channel at a fixed location in frequency, time, and space can be viewed as an equivalent AWGN channel: $Y' := Y/h = X + N/h$. The Shannon capacity of this channel gives the maximum data rate at which information can be transmitted over the channel with arbitrarily small probability of error.[48] For a fixed $H_z = h$, and assuming perfect RCSI, i.e., $\hat{H}_R = h$, the capacity is

$$C_{RCSI}(h) = \log\left(1 + \frac{|h|^2 E_s}{N_0}\right). \tag{2.116}$$

It is clear that for each channel realization traditional Shannon theory allows a strong characterization of the wireless channel.

However, wireless systems operate over a large number of positions in frequency, time, and space, and the different channels seen at these different positions make the performance random. Communication engineers have resorted to modeling the gain, H_z, as a random variable. In the traditional sense, a wireless channel might not have a capacity as it is possible that positions exist in frequency, time, and space where $H_z = 0$ and reliable communication cannot be supported. Multiple characterizations of fading channel capacity have appeared, and here the two most widely used ones will be treated: average capacity and outage capacity. The channel capacity $C_{RSI}(H_z)$ can be viewed as a random variable. The average capacity is defined as average of this random variable. Assuming that only the receiver knows H_z perfectly, this average capacity can be written as

$$C_{ave,RCSI} = E_{H_z}[C_{RCSI}(H_z)] = \int_{-\infty}^{\infty} C_{RCSI}(h) f_{H_z}(h) dh. \tag{2.117}$$

In wireless communications the most commonly used fading channel model is Rayleigh fading. Figure 2.31 gives an example of the average capacity of a Rayleigh flat-fading channel. Examining this figure it is apparent that for Rayleigh fading the average capacity is proportional to $\log_2(1 + SNR)$, where SNR refers to the average signal-to-noise-ratio over all possible channel realizations. As a result, the average capacity in a wireless channel behaves much like the average capacity in a traditional wireline channel.

An important difference from the wireline channel is seen in the outage capacity. The motivation of defining outage probability/outage capacity is to answer the following question: how often is a fading channel bad enough to not support communication of a certain rate,

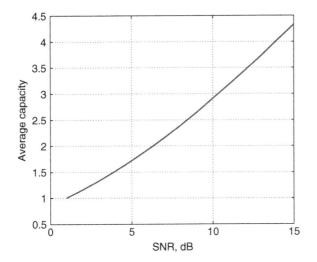

Figure 2.31   An example of the average capacity of a Rayleigh flat-fading SISO channel.

or in a dual way, what is the maximum achievable data rate given the probability that the fading channel cannot support this rate? The *outage probability* is defined as, for a given R,

$$P_{out}(R) = \Pr(C_{RCSI}(H_z) \leq R),\qquad(2.118)$$

where $C_{RCSI}(H_z)$ is a random variable that is a function of the random variable H_z. For Rayleigh flat-fading channels the outage probability can be calculated as (also see Ozarow, Shamai and Wyner[36] for a detailed treatment)

$$P_{out}(R) = 1 - \exp\left(-SNR^{-1}(2^R - 1)\right),\qquad(2.119)$$

where *SNR* is again the average signal-to-noise ratio and the unit of R is *bits/s/Hz*. An example of the outage probability for $R = 2\,bits/s/Hz$ is plotted in Figure 2.32.

The outage capacity can be defined by identifying an acceptable outage probability and then finding the information rate at which this outage probability is achieved for a given channel. This capacity can then easily be obtained from an outage probability characterization. For example, the 10% outage capacity is defined as

$$C_{0.1} = \sup\{R : \Pr\left(R > C_{RCSI}(H_R)\right) \leq 0.1\}.\qquad(2.120)$$

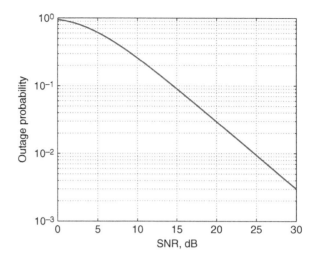

Figure 2.32 An example of the outage probability for a Rayleigh flat-fading SISO channel.

For the SISO Rayleigh channel it is easy to see the outage probability scales as $P_{out}(R) \propto SNR^{-1}$. This implies an interesting characteristic for SISO wireless communications in that the average capacity grows in a manner much like that of traditional wireline communications, logrithmic in SNR, but the outage probability only goes down with the inverse of the SNR. All users of a wireless mobile telecommunication system have experienced the frustration of having horrible reception at a location in space or time and moving slightly or waiting a short period of time and having good reception restored. This is the curse of SISO wireless systems and why the quality of service in wireless is not congruent to wireline systems.

2.7.2 Capacity of the SISO Varying Channel

Many engineers noted that channels vary in time and/or frequency and explored if this variation was the solution to the reliability issue in wireless communication. Communication theorists found many techniques to improve the reliability of wireless communications by exploiting the variability in frequency and time. Later information theory tools developed by Verdú and Han[59] allowed a nice theoretical characterization of these systems. Essentially, Verdú and Han's theory implies that if the variations in the channel in time or frequency are ergodic, then the Shannon capacity of this varying channel will converge to the average capacity. In the literature the average capacity is often denoted the *ergodic capacity* to emphasize that when channels have exploitable variations, a Shannon capacity can be defined and is equal to the ensemble average capacity of the TF/FF multipath channel. These channel variations can be exploited in a powerful way in that reliable communication can always be supported on a wireless channel. The important idea here is that if the channel gain is small at a particular point in frequency or time, this will not prevent reliable communication, as other points in frequency or time will have a large channel gain and permit the information to be reliably communicated. These ideas were the motivation for many powerful frequency and/or time domain coding techniques.

2.7.3   Capacity of the RSV/TF/FF Channel

In the second age of wireless communication, diversity was viewed as an effective solution to fading. As mentioned earlier, time and frequency diversity were extensively used in the form of channel coding and frequency hopping, respectively. Interleaving was possibly done prior to this to guarantee diminishing correlation between gains affecting adjacent symbols. These techniques improved the reliability of communication significantly. Another technique described earlier was receiver-space diversity. The RSV/TF/FF model well accounts for the channel in this case. The capacity of such RSV/TF/FF can be viewed as a generalization of the results of Section 2.7.1 to an arbitrary number of receive antennas L_r. From (2.111) and (2.112), it can be shown that the average capacity of such a channel for perfect RCSI and in the absence of TCSI is

$$C_{ave,RCSI} = E_{\vec{H}_z}\left[\log_2\left(1 + \frac{E_s}{N_0}\vec{H}_z^\dagger\vec{H}_z\right)\right] = E_{\vec{H}_z}\left[\log_2\left(1 + \frac{E_s}{N_0}\sum_{n=1}^{L_r}|H_z^{(n)}|^2\right)\right],$$

(2.121)

where † denotes the complex conjugate transpose. $C_{ave,\,RCSI}$ is plotted in Figure 2.33(a) for the case when all channel gains are independent and zero mean Gaussian random variables (Rayleigh channels).

The question becomes: what is the change that multiple receive antennas brought about compared to the SISO case? By intuition, since L_r scales the total received SNR, an increase in L_r is analogous

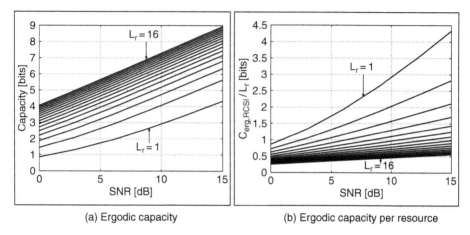

(a) Ergodic capacity   (b) Ergodic capacity per resource

Figure 2.33 Ergodic capacity and capacity per resource of a frequency-flat Rayleigh fading channel with 1 transmit antenna and L_r receive antennas.

to increasing the transmit power in the SISO case. In the high SNR regime, the strong law of large numbers[21] can be used to show that the ergodic capacity increases logarithmically with L_r. This behavior is also depicted in Figure 2.33(a), which is a plot of the average capacity versus the average SNR per receive antenna.[†] To tackle this idea from another direction, it is interesting to consider the notion of average capacity per resource. Each added antenna is a resource that raises the whole cost of the system. So, it becomes highly desirable that this resource is efficiently used to increase the data rate. We define the capacity per resource as

$$\tilde{C} = \frac{C_{ave,\ RCSI}}{L_r} \tag{2.122}$$

and plot $\tilde{C}$ versus SNR in Figure 2.33(b) for $L_r = 1, \ldots, 16$. It is clear that

$$\tilde{C} \propto \frac{\log(L_r)}{L_r} \tag{2.123}$$

which decreases by increasing L_r. This means that the more receive antennas employed, the less resource efficient the communication becomes as each antenna will contribute less to the overall possible data rate. In the limit when the number of receive antennas becomes

---

[†] Unless otherwise noted, SNR denotes the average signal-to-noise ratio per receive antenna.

very big, the capacity per resource approaches zero. Since there is a diminishing return of data rate with resources, the typical number of receive antennas used for receive diversity in a pre-1990 wireless system is $L_r \leq 4$.

The outage probability, on the other hand, is expressed as

$$P_{out}(R) = \Pr\left[C\left(\vec{H}\right) < R\right] = \Pr\left[\sum_{i=1}^{L_r} |h_i|^2 < \frac{2^R - 1}{\text{SNR}}\right] = \Pr\left[\chi < \chi_0\right] = F_\chi(\chi_0),$$

(2.124)

where $\chi = \sum_{i=1}^{L_r} |h_i|^2$ is a central chi-square random variable with $2L_r$ degrees of freedom[42] whose component Gaussians have zero mean and variance 0.5 and $\chi_0 = \frac{2^R - 1}{\text{SNR}}$. From Figure 2.34 it becomes clear that increasing L_r makes the outage probability decay much faster than the SISO case. However, this improvement becomes less significant as L_r increases. This arises from the fact that increasing L_r has the effect of shifting the trunk of the PDF curve of χ to the right.[42] Thus, the tail of the curve lying in the interval $[0, \chi_0]$ (i.e., for a specific rate R) becomes less sensitive to L_r. In the high SNR regime, the outage probability can be approximated to behave as

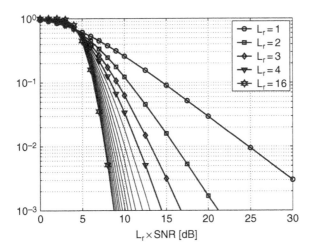

Figure 2.34   Probability of outage $\Pr\left[C\left(\vec{H}\right) < 2\right]$ of a Rayleigh fading channel with one transmit antenna and L_r receive antennas.

$$P_{out}(R) \propto \left(\frac{L_r \times \text{SNR}}{R}\right)^{-L_r} \tag{2.125}$$

This is depicted in Figure 2.34 for $R = 2$. Note that in Figure 2.34 the outage probability is plotted versus the average received SNR (i.e., $L_r \times$ SNR). The insensitivity of the outage probability to a further increase in L_r after a certain value (4 in Figure 2.34) is where this second age of wireless becomes unable to yield further fruit. In other words, given the same amount of received SNR, employing an additional antenna does not yield significant improvement in performance. Therefore, it would be less efficient to invest on receiver spatial diversity beyond a certain number of receive antennas.

The conclusion about receiver diversity is that

- It significantly improves the outage probability – although this improvement diminishes for higher L_r.

- The capacity per resource drops with larger L_r.

Consequently, in the second age of wireless, frequency, time, and space diversity were exploited to as great a degree as economically feasible. The diversity helped "stabilize" performance much more than it helped increase the information rate.

2.7.4 MIMO Capacity

In the early 1990s Telatar[54], Foschini and Gans[20] studied the capacity of a wireless channel equipped with multiple transmit and multiple receive antennas, which is also referred to as a Multiple-Input Multiple-Output (MIMO) channel. One of the most important results they suggested is that in a MIMO system one can achieve the following two advantages simultaneously:

- The spectral efficiency of such a system can grow **linearly** with the smallest number of transmit and receive antennas. More precisely, the capacity of a channel with L_t transmit, L_r receive antennas, and IID Rayleigh fading gains between each antenna pair in the high SNR regime can be approximated as $C_{RCSI} \approx \min(L_t, L_r) \log \text{SNR}$.

- $\max(L_t, L_r)$ levels of diversity can be achieved in the medium SNR regime.

This result has a fundamental impact on modern wireless system design. Starting from 1990s, there has been an explosively increasing demand of multimedia service within the limited bandwidth. At the same time, spectral resources have become more and more precious. Telatar[54] and Foschini and Gans[20] pointed out that the spectral efficiency can be greatly increased and the outage performance can be improved simultaneously by adding antennas at both sides of the communication link. This is very attractive because comparing with the time and frequency domain resources, the space domain resources are almost "infinite" and with respect to spectral efficiency are "for free". Following up on Telatar[54] and Foschini and Gans[20] there has been intense research and development efforts on MIMO techniques in both academia and industry during the past decade.

This section will focus on the capacity aspect of MIMO systems. The readers who are interested in the code design and signal processing aspects can refer to Larsson and Stoica[30] and Paulraj, Nabar and Gore.[40] By considering a single-user, point-to-point wireless channel with L_t transmit and L_r receive antennas, and assuming that the channel is frequency-flat fading, the following discrete-time model can be used:

$$\vec{Y} = \mathbf{H}_z \vec{X} + \vec{N}, \qquad (2.126)$$

where $\vec{X} = [X_0, X_1, \ldots, X_{L_t-1}]^T$ is the L_t-dimensional complex transmitted vector, $\vec{N}$ is the L_r-dimensional additive white circularly symmetric complex Gaussian noise vector with zero mean and covariance matrix $E[\vec{N}\vec{N}^\dagger] = N_0 \mathbf{I}_{L_r}$, and $\vec{Y}$ is the L_r-dimensional received vector. The channel is represented by the $L_r \times L_t$ matrix $\mathbf{H}_z = \left[H_z^{(m,n)} \right]$, where $H_z^{(m,n)}$ denotes the complex channel gain between transmit antenna n and receive antenna m which is assumed to be IID Gaussian with zero mean and unit variance. The total transmission power is assumed to be E_s, regardless of L_t.

We first investigate the average capacity of a MIMO channel. Assuming the channel is constant, the receiver knows $\mathbf{H}_z$ perfectly, and the transmitter only knows the distribution of $\mathbf{H}_z$ correctly, the capacity for a realization of the channel $\mathbf{H}_z = \mathbf{h}$ is given as

$$C_{RCSI}(\mathbf{h}) = \log \det \left(\mathbf{I}_{L_r} + \frac{E_s}{L_t N_0} \mathbf{h} \mathbf{h}^\dagger \right). \qquad (2.127)$$

Similar to the discussion in Sections 2.7.1 and 2.7.3 here both average and outage capacities in MIMO channels are considered. The average capacity can be calculated by finding the expectation over $\mathbf{H}_z$ from (2.127), i.e.,

$$C_{ave,RCSI} = E_{\mathbf{H}_z}\left[\log\det\left(\mathbf{I}_{L_r} + \frac{E_s}{L_t N_0}\mathbf{H}_z\mathbf{H}_z^\dagger\right)\right].\qquad(2.128)$$

By using random matrix theory[55] this formula can be further evaluated in an integral form (Telatar[54], §4.2). Figure 2.35(a) gives an example of how average capacity increases as a function of the number of antennas. A very interesting result is that the average capacity scales **linearly** with $\min(L_t, L_r)$. Compared with the results in Section 2.7.3, it is clear how multiple transmit antennas further increase the capacity. In fact, having multiple antennas on one side of the link can only increase the SNR, but having multiple antennas on both sides of the link allow parallel spatial channels to be formed if the scattering is rich enough.

A similar approach to that in Section 2.7.3 is to investigate the average capacity per resource. Here the total number of antennas is $L_t + L_r$, so the average capacity per resource is

$$\tilde{C} = \frac{C_{ave,RCSI}}{L_t + L_r}.\qquad(2.129)$$

Figure 2.35(b) shows an example of the average capacity per total antennas. It can be noticed that unlike the results in Section 2.7.3, here $\tilde{C}$ is slightly increased by using more transmit/receive antennas.

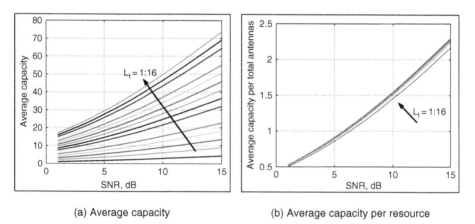

(a) Average capacity   (b) Average capacity per resource

Figure 2.35 Average capacity and capacity per resource of MIMO Rayleigh fading channels, $L_t = L_r$.

Regarding the outage capacity, Section 2.7.3 concluded that for multiple receive antennas the outage probability scales as

$$P_{out}(R) \propto \left(\frac{L_r \text{SNR}}{R}\right)^{-L_r},\qquad (2.130)$$

and receive diversity does not add much to achieve throughput but does greatly increase the reliability. Research[20,54] has revealed that MIMO gives both increased capacity and improved reliability simultaneously. At medium SNR the MIMO channel can be decomposed into $\min(L_t, L_r)$ parallel channels with $\max(L_t, L_r)$ levels of diversity.[‡] An example of the outage probability of MIMO Rayleigh fading channel when $R = L_t\, bits/s/Hz$ is plotted in Figure 2.36. The most important and interesting observation from this figure is that even while the capacity increases linearly as the number of transmit/receive antennas (see Figure 2.35), the outage probability decreases proportional to SNR^{-L_r}. Consequently, MIMO is able to provide both increased throughput and improved fidelity simultaneously.

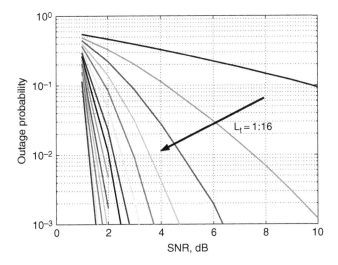

Figure 2.36 An example of the outage probability for MIMO Rayleigh flat-fading channels when $R = L_t\, bits/s/Hz$.

[‡] In the high SNR regime Zheng and Tse[64] showed that there is a fundamental tradeoff between the number of parallel channels and the achieved level of diversity.

In conclusion, a brief introduction to MIMO shows that this technique allows communication at a higher rate with a greater reliability. The remarkable capacity gain is especially exciting for the owners of wireless spectrum since the spectrum is quite limited and very expensive. MIMO makes it possible for the spectrum owners to transmit more data within the same bandwidth, at the expense of adding more Radio Frequency (RF) chains and radios. However, it is usually much cheaper to build more expensive radios than it is to buy more wireless spectrum. This can basically explain why MIMO has been such a hot topic in both academia and industry since 1995. These exciting MIMO techniques provide a possible way to increase the spectral efficiency of wireless communications.

2.8   Conclusion

This chapter has presented a brief introduction to point-to-point wireless communication. Communication theory has been driven by the pioneering work of Claude Shannon and his idea of a channel capacity. On channels that are not highly random (e.g., free space wireless or wireline), near capacity performance is achievable in a complexity that is implementable with modern technology. Multipath wireless communications produce channels that vary as a function of frequency, time, and space. This variability is what makes reliable communication over wireless multipath channels more difficult to achieve. This chapter has taken physical models of wireless propagation and communication waveforms and produced mathematical models of wireless systems that allow insight to be drawn about the evolution of modern wireless communication systems and why MIMO wireless communication has seen a significant spike in interest recently by the practitioners of wireless communication system design.

References

1. M. Abramowitz and I. E. Stegun (eds). *Handbook of Mathematical Functions*. U. S. Department of Commerce, Washington, DC, 1972.
2. S. M. Alamouti. A simple transmit diversity technique for wireless communications. *IEEE Trans. Select Areas Commun.*, 16(8):1451–1458, October 1998.

3. E. A. Armstrong. A method of reducing disturbances in radio signaling by a system of frequency modulation. In *IRE Conference*, New York, November 1935.

4. T. Aulin. A modified model for the fading signal at the mobile radio channel. *IEEE Trans. Veh. Tech.*, 28(3):182–202, August 1979.

5. J.-C. Belfiore, G. Rekaya, and E. Viterbo. The Golden code: A 2×2 full-rate space-time code with nonvanishing determinants. *IEEE Trans. Inform. Theory*, 51:1432–1436, April 2005.

6. P. Bello. Characterization of randomly time-variant linear channels. *IEEE Trans. Commun. Syst.*, CS-11:360–393, December 1963.

7. S. Benedetto and E. Biglieri. *Principles of Digital Transmission*. Kluwer Academic Press, New York, 1999.

8. E. Biglieri, D. Divaslar, P. J. McLane, and M. K. Simon. *Introduction to Trellis-Coded Modulations with Applications*. MacMillan, New York, 1991.

9. E. Biglieri, J. Proakis, and S. Shamai. Fading channels: information-theoretic and communications aspects. *IEEE Trans. Info. Theory*, 44:2619–2692, October 1998.

10. W. R. Braun and U. Dersch. A physical mobile radio channel model. *IEEE Trans. Veh. Tech.*, VT-40(2):472–482, May 1991.

11. D. W. Browne and M. P. Fitz. Field tests to measure the space-time characteristics of narrowband wireless channels in multiple-antenna systems. In *IEEE Intl. Conf. Commun.*, Helsinki, Finland, June 2001.

12. J. R. Carson. *Electric Circuit Theory and Operational Calculus*. McGraw-Hill, New York, 1926.

13. J. Cavers. *Mobile Channel Characteristics*. Kluwer Academic Press, Boston, 2000.

14. J. K. Cavers. An analysis of pilot symbol assisted modulation for Rayleigh fading channels. *IEEE Trans. Veh. Tech.*, 40:686–693, November 1991.

15. R. W. Chang. Synthesis of band-limited orthogonal signal for multichannel data transmission. *Bell Syst. Tech. J.*, 45:1775–1796, December 1966.

16. R. H. Clarke. A statistical theory for mobile radio reception. *Bell Syst. Tech. J.*, 47:957–1000, 1968.

17. T. M. Cover and J. A. Thomas. *Elements of Information Theory*. Wiley Interscience, New York, 1992.

18. G. D. Durgin. *Space-Time Wireless Channels*. Prentice Hall, Upper Saddle River, NJ, 2003.

19. M. P. Fitz, J. Grimm, and S. Siwamogsatham. A new view of performance analysis techniques in correlated Rayleigh fading. In *Wireless Commun. and Netw. Conf.*, volume 1, September 1999.

20. G. J. Foschini and M. J. Gans. On the limits of wireless communications in a fading environment when using multiple antennas. *Wireless Personal Commun.*, 6:311–335, March 1998.

21. R. Gallager. *Discrete Stochastic Processes*. Kluwer Academic, Dordrecht/Norwell, MA, 2001.

22. A. Goldsmith and P Varaiya. Capacity of fading channels with channel side information. *IEEE Trans. Info. Theory*, 43(6):1986–1992, November 1997.

23. J.-C. Guey, M. P. Fitz, M. R. Bell, and W.-Y. Kuo. Signal design for transmitter diversity wireless communication systems over Rayleigh fading channels. In *Proc. Vehicular Technology Conf.*, Atlanta, GA, April 1996.

24. J.-C. Guey, M. P. Fitz, M. R. Bell, and W.-Y. Kuo. Signal design for transmitter diversity wireless communication systems over Rayleigh fading channels. *IEEE Trans. Commun.*, 47:527–537, April 1999.

25. R. V. L. Hartley. Transmission of information. *Bell Syst. Tech. J.*, 7:535–563, July 1928.

26. B. Hassibi and M. Hochwald. High-rate codes that are linear in space and time. *IEEE Trans. Info. Theory*, 48:1804–1824, July 2002.

27. S. Haykin. *Communications Systems*. John Wiley & Sons, New York, 1983.

28. J. Heiskala and J. Terry. *OFDM Wireless LANs: A Theoretical and Practical Guide*. Sams Publishing, Indianapolis, 2001.

29. W. C. Jakes. *Microwave Mobile Communications*. Wiley, New York, 1974.

30. E. Larsson and P. Stoica. *Space-Time Block Coding For Wireless Communications*. Cambridge Univ. Press, Cambridge, UK, 2003.

31. A. Leon-Garcia. *Probability and Random Processes for Electrical Enginering*. Addison -Wesley, New York, 1989.

32. S. Lin and D. Costello. *Error Control Coding*. Prentice Hall, Englewood Cliffs, NJ, 2004.

33. Y. Z. Mohasseb and M. P. Fitz. A 3D spatio temporal simulation model for wireless channels. *IEEE J. Select. Areas Commun.*, 20(6):1193–1203, August 2002.

34. M. Morelli and U. Mengali. A comparison of pilot-aided channel estimation methods for OFDM systems. *IEEE Trans. Signal Processing*, 49:3065–3073, December 2001.

35. H. Nyquist. Certan topics in telegraph transmission theory. *AIEE Trans.*, 47:617–644, 1928.

36. L. H. Ozarow, S. Shamai, and A. D. Wyner. Information theoretic considerations for cellular mobile radio. *IEEE Trans. Veh. Tech.*, 43:359–378, May 1994.

37. J. D. Parsons. *The Mobile Radio Propagation Channel*. Halsted Press, New York, 1992.

38. J. D. Parsons and M. D. Turkmani. Characterization of mobile radio signals: Base station crosscorrelation. *IEE Proc.-I*, 138(6):557–565, December 1991.

39. J. D. Parsons and M. D. Turkmani. Characterization of mobile radio signals: Model description. *IEE Proc.-I*, 138(6):459–556, Decemer 1991.

40. A. Paulraj, R. Nabar, and D. Gore. *Introduction to Space-Time Wireless Communications*. Cambridge Univ. Press, Cambridge, UK, May 2003.

41. H. V. Poor. *An Introduction to Signal Detection and Estimation*. Springer-Verlag, New York, 1988.

42. J. G. Proakis. *Digital Communications*. McGraw-Hill, New York, 1989.

43. J. G. Proakis and M. Salehi. *Communications System Engineering*. Prentice Hall, Englewood Cliffs, NJ, 1994.

44. T.S. Rappaport. *Wireless Communications*. Prentice Hall, Englewood Cliffs, NJ, 2002.

45. S. O. Rice. Mathematical analysis of random noise. *Bell Syst. Tech. J.*, 24:96–157, 1945.

46. M. Samuel, C. Pietsch, and J. Lindner. Linear spreading space-time codes: System model and optimization criterion. In *PIMRC*, Berlin, Germany, September 2005.

47. M. Schwartz, W. R. Bennett, and S. Stein. *Communication Systems and Techniques*. McGraw-Hill, New York, 1966.

48. C. Shannon. A mathematical theory of communication. *Bell Syst. Tech. J.*, 1948.

49. C. E. Shannon. Channels with side information at the transmitter. *IBM J. Res. Develop.*, 2:289–293, October 1958.

50. R. Steele. *Mobile Radio Communications*. IEEE Press, New York, 1994.

51. G. L. Stüber. *Principles of Mobile Communication*. Kluwer Academic Publishers, Boston, 1996.

52. V. Tarokh, H. Jafarkhani, and A. R. Calderbank. Space-time block codes from orthogonal designs. *IEEE Trans. Info. Theory*, 45:1456–1466, July 1999.

53. V. Tarokh, N. Seshadri, and A. R. Calderbank. Space-time codes for high data rate wireless communications: Performance criterion and code construction. *IEEE Trans. Info. Theory*, 44:744–765, March 1998.

54. E. Telatar. Capacity of multi-antenna Gaussian channels. *AT&T-Bell Labs Tech. Rep.*, June 1995.

55. A. Tulino and S. Verdú. Random matrix theory and wireless communications. *Foundations Trends Commun. Info. Theory*, vol. 1, June 2004.

56. G. L. Turin et al. A statistical model of urban multipath propagation. *IEEE Trans. Veh. Tech.*, VT-21:1–9, February 1972.

57. H. L. Van Trees. *Detection, Estimation, and Modulation Theory*. John Wiley & Sons, New York, 1968.

58. S. Verdú. *Multiuser Detection*. Cambridge Press, Cambridge, UK, 1998.

59. S. Verdú and T. S. Han. A general formula for capacity. *IEEE Trans. Info. Theory*, IT-40:1147–1157, July 1994.

60. A. J. Viterbi. *CDMA: Principles of Spread Spectrum Communication*. Addison-Wesley, Reading, MA, 1995.

61. S. B. Wicker. *Error Control Systems for Digital Communications and Storage*. Prentice Hall, Upper Saddle River, NJ, 1995.

62. N. Wiener. *The Extrapolation, Interpolation, and Smoothing of Stationary Time Series with Engineering Applications*. Wiley, New York, 1949.

63. J. Wu and W. Wu. A comparative study of robust channel estimators for OFDM systems. In *Proc. IEEE Intl. Conf. Commun. Tech.*, Beijing, China, April 2003.

64. L. Zheng and D. N. C. Tse. Diversity and multiplexing: a fundamental tradeoff in multiple-antenna channels. *IEEE Trans. Info. Theory*, 49:1073–1096, May 2003.

3

Handset Communication Antennas, Including Human Interactions

Y. Rahmat-Samii and Zhan Li

3.1 Introduction

Since cellular phone service was introduced commercially, the handset is no longer a voice-only mobile unit as it had been in the past. The market demands not only smaller handsets, but also multi-function devices. Furthermore, electromagnetic (EM) radiation has always been a concern to the general public. In this chapter we investigate the Planar Inverted "F" Antenna (PIFA) designs for handsets, multiple antennas integration, and the antenna interaction with the human head. Figure 3.1 shows the overview of this chapter.

3.1.1 Mobile Communication Systems

In general, antennas for handsets have been required to be small, lightweight, and low profile and to have an omni-directional radiation pattern in the horizontal plane. However, with the evolution of mobile communication systems, antenna technology has also progressed, and the challenge for handset antennas goes much further.

In the current handset industry, there are many cellular communication systems, as listed in Table 3.1. In Europe and Asia 900/1800 MHz has been used for the Global System for Mobile Communication (GSM) system, while 800/1900 MHz is allocated in the United States for both

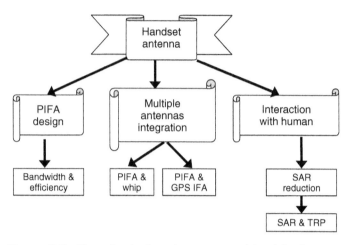

Figure 3.1 Flowchart of topics presented in this chapter.

Table 3.1 Frequency bands of cellular services and non-cellular applications.

Cellular Service		Non-Cellular Applications	
	Frequency		Frequency
GSM 900/1800 (Global System for Mobile Communication)	890–960 MHz, 1710–1880 MHz	WLAN (Wireless Local Area Networks)	2.4–2.484 GHz
GSM 800/1900	824–894 MHz, 1850–1990 MHz	BT (Bluetooth)	2.4 GHz
CDMA 800/1900 (Code Division Multiple Access)	824–894 MHz, 1850–1990 MHz	GPS (Global Positioning System)	1575.42 MHz
DCS (Digital Communication System)	1710–1880 MHz	Digital TV	470–770 MHz
UMTS (Universal Mobile Telecommunication System)	1920–2170 MHz	FM Radio	88–108 MHz
AMPS (Advanced Mobile Phone Service)	824–894 MHz	RFID (Radio Frequency Identification)	125–135 kHz, 13.56, 869.5, 915 MHz, 2.4 GHz

GSM and Code Division Multiple Access (CDMA) systems. In this chapter, antennas are designed for CDMA 800/1900 MHz handsets.

As mentioned earlier, miniaturization is one of the antenna design targets for handsets. However, the shrinkage of the handset antenna makes it more and more challenging to achieve the adequate bandwidth and efficiency.

On the other hand, non-cellular antennas have received more and more attention as new applications (listed in Table 3.1) have been integrated into the handset. For example, Enhanced 911 (E911) service for locating wireless users has been mandatory in the U.S. handset market since the year 2003 by the Federal Communication Commission (FCC).[1,2] In some handsets E911 has been accomplished by using Global Positioning System (GPS) technology, which requires a small GPS antenna to be integrated into the handset.[3] Unlike the ordinary GPS service, which provides a map on the screen to identify the current location, E911 service enables the wireless carriers, dispatchers, and local phone companies to pinpoint the mobile callers who are on the emergency 911 calls. Bluetooth (BT) has also become quite popular because of its convenience to users, which again requires the handset to have a built-in BT antenna. Recently, the FCC also opened part of the frequency spectrum for digital TV (also referred to as video broadcasting) applications, and as a result, it has been projected that many handset manufactures will integrate the TV function into new handsets in the near future.[4] Radio Frequency Identification (RFID) has been around since World War II. Recent integrated circuit technology developments have opened the door to many new applications of RFID.[5,6] Handset manufactures have begun to integrate this technology into the phone as well, which requires a small RFID tag antenna.

Due to the huge appetite of the public for handsets, the number of end users is growing exponentially, which necessitates the increase of the mobile system capacity in the near future. It is shown that by the use of spatial diversity (multiple antennas) in the mobile unit, a significant increase in system capacity can be achieved.[7] Another advantage of antenna diversity is that the detrimental effects of signal fades in a multi-path environment can be mitigated. This type of fading occurs when multiple replicas of the desired signal arrive at the receiver over different paths, thus having different relative amplitudes and phases.[8] A typical diversity handset requires two separate cellular antennas to be integrated into one unit to provide separate channels.

To meet the multi-function capability of the handsets, multiple antennas integration has become the trend for handset development. However, it has also been projected that multiple antennas integration will face a major challenge, namely, the coupling between the various antennas due to the close proximity from the small size of the handset compared to the wavelengths involved.

3.1.2   Antenna Designs for Handsets

Figure 3.2 shows some prototype antennas studied in this chapter. PIFAs have been used in many models of handsets on the market. They have drawn a lot of attention due to their compact size and low cost.[9-13] Compared to the traditional stubby antennas, PIFAs are also concealable within the housing of the handset, which makes them less likely to be damaged. The research in this chapter will start from the PIFA design. In fact, compared to PIFAs, stubby antennas depend much less on the ground plane, which makes them relatively easier to design. However, at the same time, stubby antennas also make the handset less attractive and desirable to the users due to their extruded location.

Whip antennas are still in use in the U.S. CDMA market. Because there are fewer base stations in a CDMA network, the handset with only a PIFA has a difficult time meeting the performance requirements from mobile carriers. Especially, when the handset with a PIFA only is put against the user's head, the gain can drop up to 3 dB. A PIFA-Whip combination has thus become a popular design, which helps to overcome this drawback. Solutions to reduce the coupling between the whip and the PIFA will be provided in this chapter.

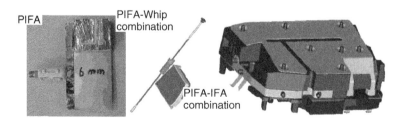

Figure 3.2 Configurations of the PIFA prototype, PIFA-Whip, and PIFA-IFA combinations.

When a GPS antenna is required to be integrated into a handset, it is usually cost effective to add a side-mounted Inverted "F" Antenna (IFA) to the PIFA module from a manufacturing point of view.[14] A ceramic antenna is another option for the GPS solution. However, it is more expensive than the integrated IFA, and it requires a fairly large amount of space on the printed wiring board (PWB) in the handset. Designing a tri-band PIFA to include the GPS band[15] may be undesirable because the system would need a tri-plexer (for a dual-band system) or an electrical Radio Frequency (RF) switch to separate the GPS signal from the cellular signals. Either a tri-plexer or a switch introduces an undesirable insertion loss of at least 0.5 dB to the GPS receiver. In addition, the use of a switch prevents the use of the cellular antenna and the GPS antenna at the same time. Quality factor (Q) value, efficiency, and polarization are usually the most important specifications for GPS antennas. But again, since the GPS frequency is so close to the 1900 MHz frequency band, the coupling between the GPS antenna and the PIFA has to be treated carefully.

BT or Wireless Local Area Networks (WLAN) antennas are relatively small and operate at higher frequencies for short ranges, which relax the problem of coupling to the cellular antennas significantly. Due to the uncertainty of the RFID standards, it is hard to focus on some particular type of RFID antennas. For example, in Europe, 13.56 MHz has been widely used for RFID applications, which requires a fairly big antenna. In the United States, 915 MHz and 2.4 GHz draw more attention. Due to the above reasons, BT/WLAN and RFID antennas are not discussed in this chapter.

3.1.3 Interaction with the Human

Because handsets operate in close proximity to users, one particular consideration involves the interaction of the EM fields with the human head, as shown in Figure 3.3. The user's influence on the antenna efficiency, radiation pattern, input impedance, and polarization[16,17] is an issue that deserves careful investigation.

Specific Absorption Rate (SAR) is one of the key considerations in antenna designs for handsets.[16] In the Unites States, the FCC established that SAR from exposure to EM radiation, as averaged over 1 g of tissue, must be lower than 1.6 mW/g (or 1.6 W/kg). Therefore, all handset models must pass SAR test limits and receive FCC certification before being sold to the general public. Since August 2000, the Cellular

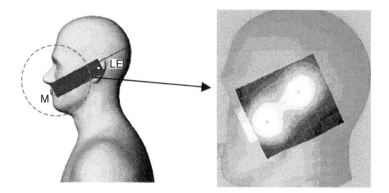

Figure 3.3 Handset antenna interaction with the human head.

Telecommunications & Internet Association (CTIA) has even required all new certified handsets to have their SAR values listed in their user manuals. In this chapter, methods to lower SAR values will be discussed. The near-field distribution will be determined to understand the power absorption characteristics within the human head. Total Radiated Power (TRP) is another important factor that affects SAR, since the CTIA has specified over-the-air performance tests for handsets, including TRP tests.[18] The TRP metric takes into account both SAR and antenna efficiency, and it correlates well with the handset field performance.

3.1.4 Objectives of this Chapter

In this chapter, practical and novel designs for handset antennas are provided and optimized and are based on some industrial requirements. Mechanical restrictions are also applied to make the designs more representative for the real handsets. Section 3.2 presents an overview of different types of handset antennas. PIFA designs start in Section 3.3, which is followed by the study of the multiple antenna solutions that combines a PIFA with other antennas. Antenna interactions with the human head are discussed in Sections 3.4 and 3.5 by also giving consideration to the TRP parameter.

This chapter focuses on handset antenna designs based on the authors' recent research endeavors and published works. Prototype handset antennas were fabricated and measured using antenna far-field ranges, and the plane cut of radiation patterns is defined in Figure 3.4. The SAR was measured with a Dosimetric Assessment System (DASY) 4 measurement system,[19] as shown in Figure 3.5.

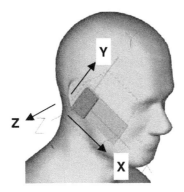

Figure 3.4 Definition of the plane cut of the radiation patterns for both the simulation and the measurement. It can also be applied to the patterns in the free space. Typically, the XZ plane is used for two-dimensional cut.

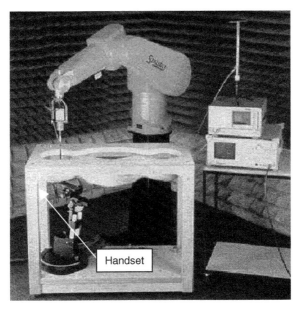

Figure 3.5 Dosimetric Assessment System (DASY) 4 SAR measurement system. The electric probe controlled by the robot can reach into the fluid (head or body stimulant) to measure the E field inside the fluid.

3.2   Overview of Popular Handset Antennas

Since the beginning of the 1990s, several development generations of handsets have appeared on the market, which is now marked by a strong competition between the handset manufacturers and by noticeably strong demand for attractive designs and technical improvements.[20] On today's market, there are three major phone factors, namely, the candy-bar handset, the sliding handset, and the clamshell handset. Internal antennas (or integrated back-mounted antennas[16]) are used mostly in the candy-bar and sliding handsets. External antennas are more popular in the clamshell handsets due to the shorter ground plane. Compared to external antennas, internal antennas have increasing demands in order to improve the industry design, but also because of mechanical advantages such as easy mass production and better handling. However, external antennas still play an important role because in some situations better performance can be achieved when the handset is put against the user's head.

Besides the cellular antennas used for voice and data services, other applications also have been added to current handsets, which will require the integration of non-cellular antennas such as the GPS antenna, BT antenna, WLAN antenna, and so on. As a consequence, a multiple antenna solution has become the trend for the handset antenna designs. Another important application, which requires multiple antenna integration, is antenna diversity. It will evolve due to the technical advancement of the Multiple-Input Multiple-Output (MIMO) systems. After a brief introduction of the RF system, this section provides an overview of different types of antennas used in current handsets.

3.2.1   RF System Introduction

In a handset, the mechanics, such as the chassis and the display, will have a great impact on the antenna performance because the mechanics is viewed as part of the antenna system. Besides the mechanics, the nature of the mobile system itself, especially the RF system section, greatly influences the ultimate antenna design. A simple RF system section in a handset is shown in Figure 3.6.

Along the receiver (RX) path, the modulated RF signal is picked up by the antenna and passed through the diplexer and duplexer to the low noise amplifier (LNA). The diplexer separates the 800 MHz

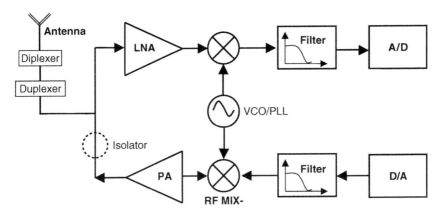

Figure 3.6 An oversimplified block diagram of a typical RF system in a handset. Many filters are skipped, and GPS or BT portions are not included.

band signal and the 1900 MHz band signal. The duplexer is used to separate the RX and the transmit (TX) paths. Along the TX path, the modulated RF signal is delivered to the antenna by the power amplifier (PA) through the diplexer and the duplexer. One major goal of the antenna design is to match both the input impedance of the LNA and the output impedance of the PA, which requires the antenna to have an adequate bandwidth to cover both the TX band and the RX band.

As addressed in Section 3.1, if a dual-band system also includes the GPS receiver, an easy way would be to integrate all the RF sections together to make a tri-band system with a tri-band antenna. Such a tri-band system is undesirable because, compared to a typical duplexer, the insertion loss of the tri-plexer will be much worse and cause more degradation to the receiver. Using an RF switch instead of a tri-plexer can improve the insertion loss. However, the RF switch creates another issue, namely, that making the phone call and retrieving the GPS location data cannot occur simultaneously.

Therefore, it is very important to understand the RF system before choosing a solution for the antenna design.

3.2.2 External Antennas

The advantage of external antennas is that the dependency on the PWB or the ground plane is greatly reduced. Typical external antennas on the market are whip antennas (monopoles) and stubby antennas. External antennas are usually designed to be retractable so that when antennas are extended, better antenna performance against the human head can be achieved. On the other hand, minimum performance has to be maintained with antennas retracted.

3.2.2.1 Whip antennas

Figure 3.7 shows the whip antenna. At the top, a plastic cap is used to pull the whip out. The bottom part will be connected to the feed point on the PWB. Conventional whip antennas are quarter wavelength (1/4 λ) monopoles.[21] Using the ground plane to create an image, the monopole unit behaves in the same way as the dipole antenna. Therefore, a gain of 2.15 dB can be achieved.[22] At the 800 MHz band, the radiation pattern is a typical donut shape, while at the 1900 MHz band, a butterfly-shape pattern is usually seen. Figure 3.8 shows the patterns of an active whip. The electrical length of the monopole determines the input impedance and the resonant frequency; however, the position of the monopole antenna in the handset does not significantly affect the impedance and the resonant frequency.

There are two common solutions for a dual-band whip antenna. First, by using a matching circuit one creates a dual-band active whip. In this case, the whip works independently in the extended position. Two reasons make this solution less attractive. One reason is that, in order to make the extended active whip work efficiently, the stubby antenna or the PIFA has to be detuned not to affect the whip. This can require a complicated mechanical design. The other reason is that the SAR value of an active whip tends to become high because the antenna radiates evenly to the front and the back of the handset.

The second dual-band whip solution is to use a parasitic whip antenna, which is electromagnetically coupled through a dual-band PIFA or a stubby antenna. Compared to the active whip solution, the parasitic whip is more sensitive to the human head. By careful design

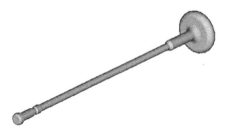

Figure 3.7   The whip antenna, plastic-coated nickel titanic, and the plastic cap for the user to extend or retract.

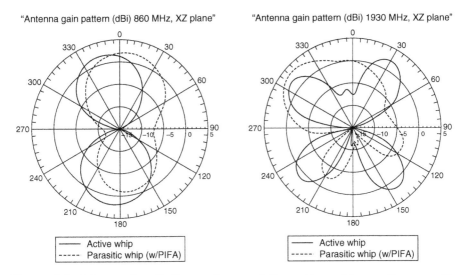

"Antenna gain pattern (dBi) 860 MHz, XZ plane"   "Antenna gain pattern (dBi) 1930 MHz, XZ plane"

Active whip
Parasitic whip (w/PIFA)

Active whip
Parasitic whip (w/PIFA)

Figure 3.8 Measured radiation patterns of the active whip vs. the parasitic whip with PIFA in the free space. The whip is 120 mm long, and the ground plane is 100 mm long. The XZ plane is defined in Figure 3.4.

of the parasitic whip, this antenna can achieve the same peak gain as the active whip. However, the radiation patterns of the parasitic whip will depend on the patterns of the PIFA or the stubby antenna.

In Figure 3.8 the dotted lines are radiation patterns of a parasitic whip coupled through the PIFA. As a consequence, these patterns are dependent on patterns of the PIFA, which have very good front-to-back ratios. Today, whip antennas are still widely used in the U.S. handset market.

3.2.2.2   Stubby antennas

Stubby antennas are another type of external antennas. They are made either of the helix or of the meander (shown in Figure 3.9). The helix is directly made of metal wire, while the meander is basically made of flex with printed wire. The meander is relatively easier and more consistent for mass production.

Stubby antennas have less dependency on the size of the ground plane than internal antennas. However, compared to internal antennas, protruded stubby antennas can be damaged more easily. Figure 3.10 shows the comparison of the radiation patterns between a PIFA and a stubby antenna. The ground plane length is fixed at 110 mm. Clearly, the PIFA has a much broader beam width at the

Figure 3.9 A stubby antenna and the meander.

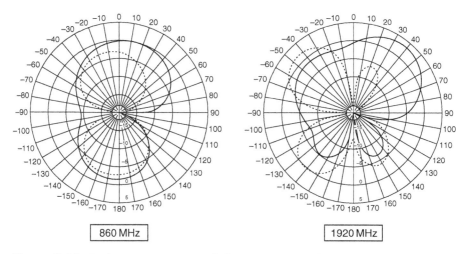

860 MHz   1920 MHz

Figure 3.10 Radiation patterns of the PIFA vs. stubby antenna at the XZ plane; the ground plane size is 110 mm long. The solid lines represent the patterns of the PIFA, while the dotted lines represent the patterns of the stubby antenna.

front, which potentially results in better performance when the handset is put against the user.

3.2.3 Internal Antennas

Because of mechanical advantages, internal antennas have been adopted by most handset manufacturers. Internal antenna solutions

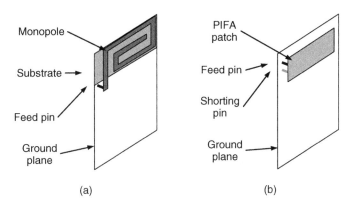

Figure 3.11 Configurations of the internal antennas: (a) The planar monopole antenna requires no ground plane underneath, while (b) the PIFA requires a solid ground plane underneath.

are based on planar antenna principles. The most common internal antennas are PIFAs[16] and folded monopoles.[13] Configurations of the two types of antennas are shown in Figure 3.11. The major difference between them is that the ground plane directly underneath the monopole is cut out, while for the case of the PIFA the ground plane underneath the patch is required, which will help block the backward RF radiation and lower the SAR values.[16]

In this chapter, the majority are of discussions are dedicated to PIFAs, including the concept of combination with the other antennas and of interaction with the human head. Novel designs of folded monopole antennas have achieved a low profile,[23] but they are more suitable to be mounted at the bottom of the handset due to SAR consideration. However, the hand will then have a substantial impact on the performance of the bottom-mounted monopole.

3.2.4   Non-Cellular Antennas

Some non-cellular applications, which require non-cellular antennas, have been introduced in Section 3.1.1. Most commonly, integrated non-cellular antennas in the current handsets are GPS antennas and BT antennas. WLAN services can share the BT antenna since the operation frequency of both WLAN and BT is the same (2.4 GHz).

IFAs and ceramic antennas, shown in Figure 3.12, are usually used as the solution for non-cellular applications because they are low profile

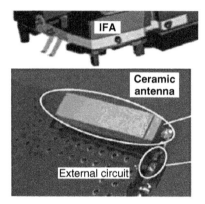

Figure 3.12 Non-cellular antennas used in handsets.

and low cost. However, depending on the application and design restrictions, different solutions will have different advantages.

For FM radio applications, basically one of the headset wires is currently used as the antenna. In terms of the wavelength, as the handset size is much smaller than the wavelength at the FM radio frequency (80~110 MHz), integration of a separate FM radio antenna is extremely challenging. Therefore, an internal FM radio antenna is rarely seen in the market to date. Digital TV or RFID applications have just started, and the standards are still under development, but it can be projected that integration of digital TV and RFID antennas will become future challenges for antenna designers.

3.2.4.1 GPS antennas

GPS is a satellite navigation system which provides specially coded satellite signals that can be received by a GPS antenna and processed in a GPS receiver, enabling the receiver to compute position, velocity, and time. The operating frequency is 1575.42 MHz. Once the handset is integrated with the GPS receiver, the user can be located upon request (E911 call).[3,24] In the current handset market, three solutions have been used for GPS service. They are tri-band cellular antennas, IFAs, and ceramic antennas. In this chapter, a PIFA-IFA combination is selected to provide a cost effective and efficient GPS solution.

3.2.4.2 BT/WLAN antennas

BT essentially is cableless networking or wireless networking. The actual means of BT working is achieved by "radio waves". Free space is the medium. Since the frequency chosen for this application is

2.4 GHz, the BT antenna is much smaller than the normal PIFA. Therefore, it is very easy for the BT antenna to fit into the handset chassis. The BT functions include file transferring between handsets, wireless headset, and communication between the handset and the BT-enabled laptop or printer.

A ceramic chip antenna is one option for BT. However, the commercialized ceramic package only gives external matching circuits to use, which limits the design ability. By designing a plated IFA[25] directly on the PWB, the length of the antenna, the feed, and the ground location can be controlled. Both of the above methods require certain space on the PWB. Designing an IFA module mechanically integrated into the handset chassis is also very simple to realize.

As mentioned before, the WLAN antenna can share the same antenna of the BT.

3.2.4.3 RFID antennas

RFID technology is an automatic way to collect product, place and time, or transaction data quickly and easily without human intervention or error. An RFID system consists of a reader that uses an antenna to transmit radio energy to interrogate a transponder (a radio tag, or RFID card). The transponder does not have a battery, but rather receives its energy from the incoming RF signal. The energy is used to extract data stored in an integrated circuit and send it to the reader, from where it can be fed to a computer for processing.[6]

Designing an efficient antenna to be integrated into the RFID tag is quite challenging. Obviously, the size and cost of such antennas are two major considerations. However, the RFID systems of today employ many competing mutually incompatible protocols which use different frequency bands, as listed in Table 3.1. This leads to research on many different antenna configurations. PIFA, IFA, printed, or folded dipole/monopole are some typical RFID tag antennas.[26,27]

3.2.5 Key Electrical Parameters in Handset Antenna Designs

In this section, definitions of some key electrical parameters, which are used for antenna designs, are briefly reviewed first.

The antenna input impedance is defined as the ratio of the voltage (V) to the current (I) at the antenna feed point:

$$Z_{in} = \frac{V}{I}|_{\text{at antenna feed point.}} \tag{3.1}$$

Normally, the input impedance is obtained by measuring the S_{11} of the antenna, which is the reflection coefficient of the antenna at the feed point:

$$S_{11} = \frac{Z_{in} - Z_o}{Z_{in} + Z_o},$$

(3.2)

where Z_o is the characteristic impedance of the feeding transmission line. For the coaxial cable, $Z_o = 50\Omega$.

The bandwidth (BW) as a percentage of the center frequency is defined in[21]

$$BW = \frac{f_u - f_l}{f_c} \times 100\%,$$

(3.3)

where $f_u, f_l,$ and f_c are the upper, lower, and center frequencies of operation. Also, f_u and f_l are usually determined by the S_{11}. For example, a standard of $-6\,dB$ return loss (S_{11}) can be used to measure the BW. From an engineering point of view, a -6 dB return loss is sufficient for an antenna to radiate fairly efficiently.

The quality factor (Q) represents the antenna losses. Usually,

$$Q \approx 1/BW.$$

(3.4)

The efficiency is another important antenna parameter, which can be defined as

$$e_0 = \frac{P_{rad}}{P_{del}},$$

(3.5)

where P_{rad} is the radiated power to the far-field region, and P_{del} is the antenna delivered power. Efficiency is a parameter that describes the antenna losses. On the other hand, directivity is often used, which is defined as "the ratio of the intensity, in a given direction, to the radiation intensity that would be obtained if the power accepted by the antenna were radiated isotropically".[21] Gain and directivity are related by Eq. (3.5).

Two major design considerations for a cellular antenna are the impedance BW and the antenna efficiency. However, as the market demands smaller handsets, the ground plane size has to shrink, which makes it very challenging to meet the required BW and efficiency at the same time. In handsets where no isolator is used between the antenna and the PA, the antenna must present more uniform impedance to the PA in order to allow the PA to operate properly.[28] In these types of handsets, additional BW is required to compensate for the detuning of the antenna by human fingers or other factors.

Since the satellites are transmitting Right-Handed Circular Polarization (RHCP) waves, the polarization of the GPS antenna in the handset front direction (away from the head in the user position) should match the satellite signal and be RHCP. Besides the polarization, the antenna efficiency is also very critical in the design because the GPS signal from the satellites is very weak due to the long distance needed to reach the user. A high Q GPS antenna is preferred, as it can reduce the out-of-band noise and thus help improve the GPS receiver sensitivity.

Compared to other applications, BT/WLAN uses a higher frequency (2.4 GHz). Therefore, antenna efficiency is the dominant parameter.

In different standards, different types of antennas can be chosen as the RFID tag antennas used in either short range (less than 2 m) or long range (10 m or longer). It is desirable to have the antenna impedance match directly to the microchip of the tag, which can be as high as 1200Ω in some products. Omni-directional or hemispherical coverage is also important.

Viewing a handset as a whole unit, which integrates multiple antennas, the coupling between each of the antennas must be optimized. SAR, which is another key parameter, must meet the FCC requirement.

3.3   Integration of Multiple Antennas

As commercial needs have expanded the functions incorporated in wireless cellular handsets, multiple antenna development in one handset has become more and more common. However, the antenna efficiency can be greatly affected by the limited isolation between each of the antennas. Some handsets have a combination of two antennas for the cellular bands for enhanced coverage and are called dual-antenna modules. The most commonly used dual-antenna modules are Stubby-Whip and PIFA-Whip combinations. For both cases, specially designed decoupling matching circuits can be used to decouple the retracted whip from the other antenna.[29] When a separate GPS antenna is integrated into the design, the coupling between the cellular antennas and the GPS antenna must be confronted as well. With the addition of a BT antenna (2.4 GHz) and a digital TV antenna (470~770 MHz), even more severe coupling problems can be expected. In the near future, to increase the network capacity, dual-receiver handsets with antenna diversity will be developed, which will also complicate the coupling issue.

In this section, a dual-band PIFA design will be introduced first. Then multiple antennas designed into a single handset will be investigated. The target is to minimize the coupling between different antennas. The combination of dual cellular antennas, PIFA-Whip, will be presented in Section 3.3.1. The combination of one cellular antenna and one non-cellular antenna, namely, a cellular PIFA and a GPS IFA, will be discussed in Section 3.3.3.

3.3.1 Dual-Band PIFA Design

The design of a PIFA comes from the traditional microstrip patch antenna.[21] In addition to the feed pin connected to the patch, another shorting pin is placed to connect the patch to the ground. Basic patch antennas usually have a limited bandwidth of about 2~4%. The patch length is approximately a half wavelength of the resonant frequency. A PIFA has improved bandwidth due to the additional shorting pin, and the patch of the PIFA is typically a quarter wavelength long, which results in a more compact profile. By cutting slots on the patch of the PIFA, or adding parasitic elements that couples to the patch, a dual-band or multi-band antenna can be achieved.

The two primary bands for U.S. mobile services are the cellular band (824~894 MHz) and the PCS band (1850~1990 MHz). A simple and efficient J-shape slot is used to design an 800/1900 MHz PIFA to cover the two user bands, as shown in Figure 3.13. The slot is cut on a 26 × 34 mm patch. The ground plane size is 78 × 34 mm, which is exactly three times the size of the patch. The substrate height is 10 mm with a dielectric constant of 2.25. A coaxial feed has been used in the simulation.

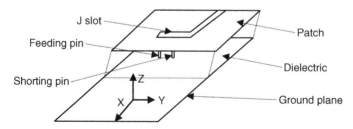

Figure 3.13 Dual-band PIFA configuration with a J-shape slot.

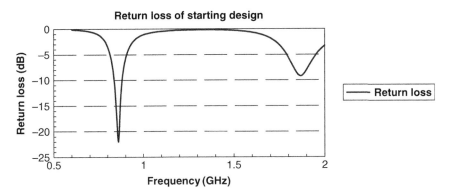

Figure 3.14 Return loss of the starting dual-band PIFA design, covering both the 800 MHz band (824~894 MHz) and the 1900 MHz band (1850~ 1990 MHz).

The return loss of the starting design is plotted in Figure 3.14. At the 800 MHz band, a bandwidth of 80 MHz is achieved using a standard of −6 dB return loss. At the 1900 MHz band, a bandwidth of 120 MHz is achieved. A –6 dB return loss is sufficient for an antenna to radiate fairly efficiently and is acceptable in practice.

The impedance BW and the antenna efficiency of a PIFA are greatly affected by the ground plane size.[30,31] However, in order to meet the market demands, the ground plane size has to shrink to fit into the smaller handsets and this causes the BW to reduce. To achieve good BW with a smaller ground plane becomes a very big challenge in the design. In addition, in the wireless industry the volume of the antenna and the material of the substrate are usually decided at the beginning of the design, which limits the design flexibility even more.

3.3.2   PIFA and Whip Antenna Combination

Since the protruded whip antenna tends to break off and increases the total handset length, novel designs such as a folded whip or a planar whip have been developed. However, these novel designs lose the advantage given by the extended whip antenna of improved performance against the head. Typically, compared to PIFAs, at the 800 MHz band, the efficiency will improve up to 3 dB with a whip when the handset is put against a human head. At the 1900 MHz band, the whip will also improve the gain by about 1~2 dB. In the weak signal area, the use of the whip can help the handset to maintain and initialize

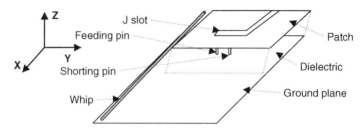

Figure 3.15   Configuration of the PIFA-Whip model at the retracted position.

phone calls. Due to all of the above, the whip antenna is still needed, and it is usually made retractable and combined together with a PIFA to keep the whole handset low profile, as shown in Figure 3.15.

However, the coupling between the PIFA and the retracted whip can significantly degrade the PIFA's radiation performance. This section focuses on decoupling the retracted whip in order to make the PIFA more efficient at a retracted situation. A matching circuit of lumped elements was used to detune the retracted whip. An LC resonant (inductance/capacity) circuit model was developed in RF ADS to assist the antenna full wave analysis in HFSS to select the optimum LC component values. At the same time, the impedance model calculated in HFSS was transferred into an RF ADS one-port network as part of the S-parameter simulation model. We will first verify the circuit model and then, based on the circuit simulation, find the optimum solution for the PIFA-Whip combination.

3.3.2.1   Coupling Between the PIFA and the Retracted Whip

In the antenna model shown in Figure 3.15, a simple and efficient J-shape slot is used to create a dual-band PIFA to cover both the 800 and the 1900 MHz bands. The slot is cut on a 26×34 mm patch. The ground plane size is 78×34 mm. The substrate height is 10 mm with a dielectric constant of 2.25. The feed and the ground pins are chosen to be in the middle of the patch in order to achieve the best impedance BW. Finally, the ground plane is extended 4 mm to the left side to accommodate the retracted whip.

First the whip bottom end was left open about 2 mm above the ground plane. Then the whip bottom end was shorted to the ground with a metal sheet. The return losses of both cases are shown in Figure 3.16. The return loss of the PIFA without the whip is also shown at the same time. It is clearly observed that when the retracted

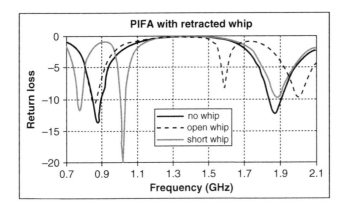

Figure 3.16   Return losses of the PIFA with retracted whip.

whip was left open, the coupling between the PIFA and the retracted whip created a destructive resonance in the 1900 MHz band. In the other case, with the retracted whip directly shorted to the ground, the coupling resonance shifted down to the 800 MHz band and mostly degraded the 800 MHz band. The radiation patterns are shown in Figures 3.17–3.19.

The peak gain and average gain are listed in Table 3.2. From the pattern plots, it is observed that at 1900 MHz, the radiation patterns of the PIFA with an open retracted whip changed dramatically from

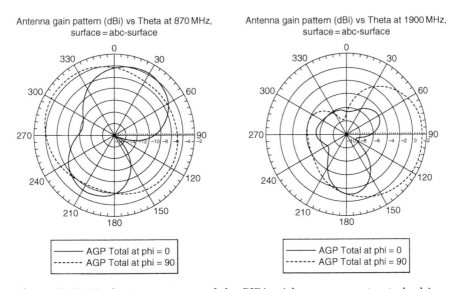

Figure 3.17   Radiation patterns of the PIFA with an open retracted whip.

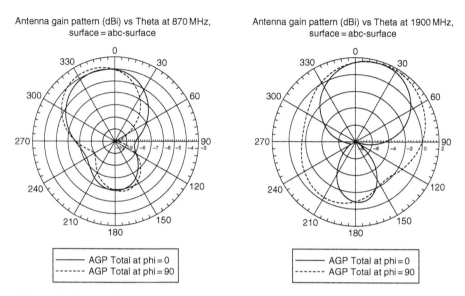

Figure 3.18 Radiation patterns of the PIFA with a shorted retracted whip.

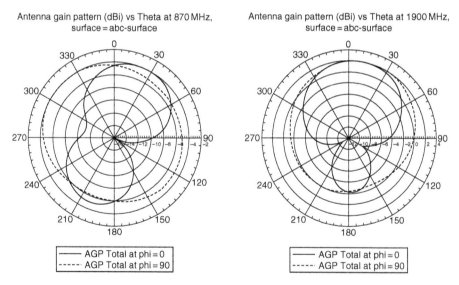

Figure 3.19 Radiation patterns of the PIFA-only case.

that of the PIFA without a whip. Not only the beam peak shift to the back (in the direction of 180°), but the average gain also dropped almost 2 dB. Similarly, the average gain of the PIFA with a shorted retracted whip at 870 MHz dropped about 1.5 dB compared to that of

Table 3.2   Gain at the H plane. (The H plane is defined as the YZ plane in Figure 3.15. The matching is discussed in Sections 3.3.2.2 and 3.3.2.3.)

	Whip Open		Whip Shorted		PIFA Only		12 nH, 2.0 pf	
Frequency (MHz)	870	1900	870	1900	870	1900	870	1900
Peak (dBi)	−3.36	1.4	−3.7	1.74	−3.88	2.17	−3.26	1.27
Average gain (dBi)	−4.9	−2.39	−6.55	−0.38	−5.1	−0.54	−4.75	−0.58

the PIFA without a whip, and the beam peak shifted to 330°. All of the above shows that the coupling between the retracted whip and the PIFA caused a negative effect.

The current distributions on the open retracted whip and the shorted retracted whip are shown in Figures 3.20–3.23. By looking at the surface current distributions, it is observed that strong coupling occurred between the whip antenna and the PIFA. It is that coupling which degrades the PIFA performance. For the case of the open retracted whip at 1900 MHz, the strength of the coupled current reaches almost 100 A/m. For the case of the shorted retracted whip at 870 MHz, the strength of the coupled current reaches almost 200 A/m. In both cases, the strong currents on the whip radiate together with the PIFA; unfortunately, the coupled currents on the retracted whip are not in phase with the feeding currents on the PIFA. As a result, the

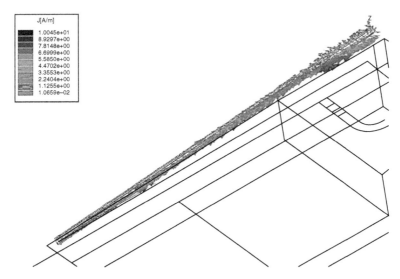

Figure 3.20   Current distribution on the open retracted whip at 870 MHz.

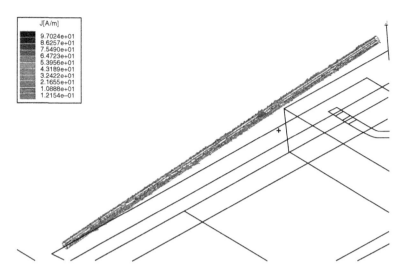

Figure 3.21 Current distribution on the open retracted whip at 1900 MHz.

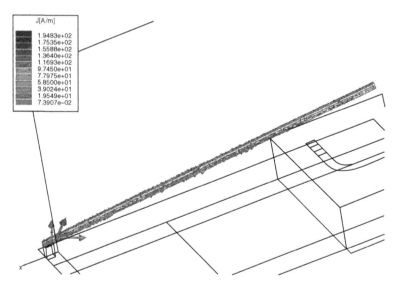

Figure 3.22 Current distribution on the shorted retracted whip at 870 MHz.

performance of the PIFA degraded significantly. In addition, the beam peaks of the radiation patterns also shifted.

It is almost impossible to control the phase of the coupled currents on the retracted whip to make it in phase with the currents on the PIFA. Therefore, in order to make the PIFA work more efficiently, one

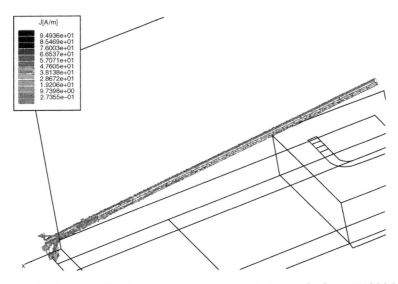

Figure 3.23 Current distribution on the shorted retracted whip at 1900 MHz.

needs to decouple the retracted whip from the PIFA. In other words, the goal is to minimize the currents on the retracted whip.

3.3.2.2 Circuit model

Since the lumped LC circuit is easy to implement, the simulation started with an inductor and a capacitor in parallel. They were added between the whip bottom and the ground plane to decouple the retracted whip. The question is how to pick the correct values for the lumped elements so that the current on the whip will be minimized. Normally, one simulation in HFSS takes about 1.5 hours. In doing the simulations with all the combinations of different LC component values requires a huge amount of computation time.

It is obvious that by adding LC elements the impedance looking into the retracted whip will change. Therefore, the coupling between the PIFA and the retracted whip will change accordingly. So first, the impedance looking into the retracted whip was calculated in HFSS and imported into RF ADS. Then the S-parameter simulation model was set up using an LC circuit as shown in Figure 3.24(a). Since the inductor and the capacitor in the HFSS model were in parallel and connected to the ground, the LC components in RF ADS were also shorted to the ground directly. However, results of the simulation in RF ADS with such a circuit model in Figure 3.24(a) did not give any

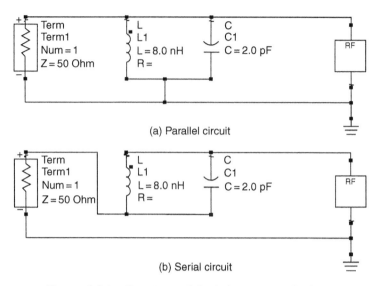

(a) Parallel circuit

(b) Serial circuit

Figure 3.24 Circuit model of the retracted whip.

hint as to how to solve this problem. That means the circuit model is not correct.

The current in the lumped elements simulated in HFSS is shown in Figure 3.25. It is clearly seen that the current flows from one element to the other without scattering into the ground area. Using this observation, the circuit model was changed into Figure 3.24(b). The major difference between the two models in Figure 3.24 is that the LC components were changed to be in series instead of directly shorted to the ground.

RF ADS simulation results of the decoupling effect are shown in Figure 3.26. In each plot, three resonances are observed; one in the 800 MHz band, one in the 1900 MHz band, and one in between. The third resonance shifts from the 800 MHz band to the 1900 MHz band as the termination of the retracted whip changes from a short to an open. The return loss of the PIFA simulated in HFSS is also shown. It is clearly observed that the three resonances in both simulations match each other very well. This proves that our circuit model in the RF ADS simulation, as shown in Figure 3.24(b), is valid to simulate the decoupling effect. The next step is to minimize the effect of those three coupling resonances shown in Figure 3.26.

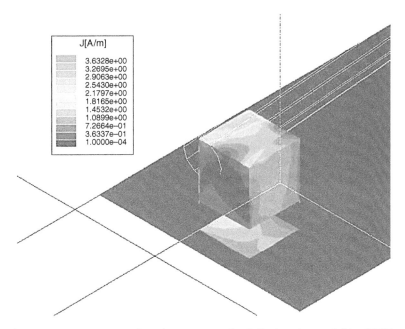

Figure 3.25 Current distribution on the LC circuit model in HFSS.

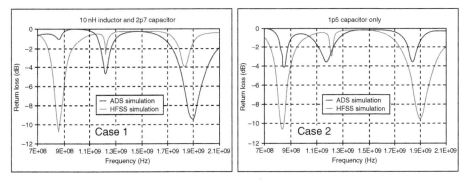

Figure 3.26 Comparison between ADS simulation and HFSS simulation. Case 1: L = 10 nH, C = 2.7 pF; Case 2: C = 1.5 pF, no L.

3.3.2.3 Decoupling the retracted whip

Since RF ADS simulation takes less than 20 seconds, the time to optimize the LC component values is significantly less than the time needed using HFSS simulation. As it has already been verified that the circuit model in Figure 3.24(b) is valid to simulate the decoupling effect, RF ADS simulation could be used to find the correct LC component values that will decouple the retracted whip from the PIFA. As a result, the gain of the PIFA could be optimized when the whip is retracted.

First, the goal was to shift the third resonance away from both the 800 and the 1900 MHz bands. After a number of searches for the LC component values, the third resonance was moved to the middle of the band, right between the 800 and the 1900 MHz bands. Then the LC values were slightly changed to minimize the other two resonances in both the 800 and 1900 MHz bands. Finally, an optimal matching circuit of a 12 nH inductor and a 2.0 pF capacitor was found. Results of both simulations are shown in Figure 3.27. The third resonance is about 1.3 GHz, so it will affect neither the 800 nor the 1900 MHz band. For the other two resonances (one at the 800 MHz band and one at the 1900 MHz band), the S_{11} level is between −2 and −3 dB, which will have a very small impact on the gain of the PIFA in those bands.

Radiation patterns of the PIFA-Whip with the optimal matching circuit are shown in Figure 3.28, which are quite close to that in Figure 3.19. The peak gain and the average gain at the H plane are listed in Table 3.2. Compared to gain values of the PIFA without the whip, only the peak gain at 1900 MHz is lower. The average gain at 1900 MHz is almost the same, and at 870 MHz, both the peak gain and the average gain are slightly better.

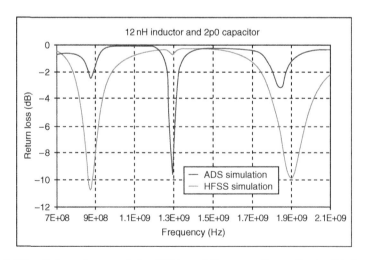

Figure 3.27   Return losses of the PIFA and the coupling effect with L = 2 nH, C = 2.0 pF.

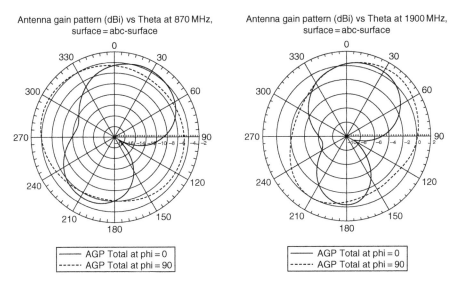

Figure 3.28   Radiation patterns of the PIFA-Whip with 12 nH and 2.0 pF.

The current distributions on the retracted whip with the optimal matching circuits are shown in Figures 3.29 and 3.30. At 870 MHz, compared to the whip-shorted case in Figure 3.22, the maximum current strength is only about 60 A/m, much lower than 200 A/m. At 1900 MHz, the maximum current strength level is below 50 A/m, only half of that of the whip-open case in Figure 3.21. It is also quite clear

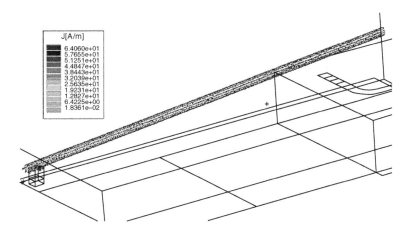

Figure 3.29   Current distribution on the retracted whip at 870 MHz, L = 12 nH, C = 2.0 pF.

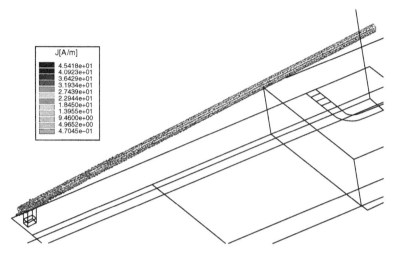

Figure 3.30 Current distribution on the retracted whip at 1.9 GHz, L = 12 nH, C = 2.0 pF.

in Figures 3.29 and 3.30, for both cases, that the current is flowing near the bottom of the retracted whip. In Figures 3.21 and 3.22, the current is flowing from the middle to the top of the whip, which is closer to the PIFA and therefore causes the strong coupling.

From all of the above, the coupling between the retracted whip and the PIFA is successfully reduced.

3.3.3 PIFA and GPS IFA Combination

This section provides a practical design of a PIFA-IFA combination module that covers the cellular bands at 800 and 1900 MHz and the GPS band. First, five antenna model candidates were simulated in Ansoft HFSS. The feed/ground locations of every model were different from each other. The total efficiencies and the IFA polarizations were compared. A prototype of the antenna model with the optimum performance was selected and then fabricated. Measurement results were used to validate the simulation results. Antenna pattern measurements of GPS in RHCP were conducted using an RHCP spiral source antenna. Then various lengths of the selected IFA were investigated. An optimum length was shown to be achievable to improve the isolation between a PIFA and an IFA in a specific frequency band.

3.3.3.1   GPS Antenna Solution

In a real handset, it is quite difficult to design a GPS antenna with a good RHCP Axial Ratio (AR) at all beam angles. Consequently, linear antennas have been used in many handset designs. A tri-band antenna provides a straightforward linearly polarized solution, but a tri-band PIFA or stubby antenna is difficult to tune. It is also very tough to meet the gain requirements in all three bands. From the system point of view, a tri-band antenna solution also requires a tri-plexer or an electrical RF switch, which will add at least a 0.5 dB insertion loss to the GPS receiver channel. It is desirable, or even required, for the GPS receiver to be able to detect GPS signals (~ -150 dBm) that are only a few decibels above the thermal noise floor. With that amount of margin, it is clear that a 0.5 dB loss is a significant amount. In addition, if an RF switch is used, the antenna can only be used to receive either the GPS or the cellular signal at any given time.

Ceramic GPS antennas have been widely used in the wireless market due to their small size and low cost (in mass production). It is primarily a linear antenna. This solution will have to occupy a certain amount of space on the PWB. Compared to other components on the PWB, the ceramic GPS antenna is still fairly large. Depending on the handset system, it might be very difficult for the GPS antenna to fit on the PWB. Another disadvantage is that since the ceramic GPS antennas have become standardized parts, antenna designers cannot modify them during the design process, but have to use an external matching network. However, the antenna's efficiency or resonant frequency will be affected by PWB changes or any other changes inside the handset housing.

IFA antennas are cost effective solutions for GPS. They can share the cellular PIFA's dielectric substrate and ground plane so that PIFA and IFA can be integrated into a single antenna module. In this way, the IFA is mainly a linear antenna, but if designed carefully, the coupling between the PIFA and IFA and placement of the IFA with respect to the ground plane could create RHCP in a certain direction. PIFA and IFA combinations will be discussed in the next section.

3.3.3.2   Location of the Feed/Ground Pins of the PIFA and the IFA

Figure 3.31 shows the IFA structure. A simple and efficient J-shaped slot was used to design a dual-band PIFA to cover the 800 and the 1900 MHz bands. The patch size was 26×34 mm. The ground plane size was

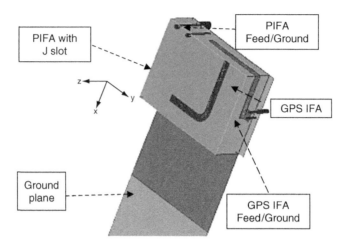

Figure 3.31 Configuration of the antenna model 1. The patch size is 26 × 34 mm. The ground plane size is 98 × 34 mm. The substrate height is 10 mm with a dielectric constant of 2.6. The GPS IFA is 4 mm below the patch. The strip width of the IFA is 2 mm. The J-slot is 2 mm wide, 21 mm long in the X direction, and 12 mm long in the Y direction.

98 × 34 mm. The substrate height was 10 mm with a dielectric constant of 2.6. The GPS IFA was put on the side of the PIFA, 4 mm below the patch. The length of the IFA was tuned to control the resonant frequency. The distance between the feed leg and the ground leg was used to control the BW.

As mentioned earlier, IFA itself is basically a linear antenna. When combined together with the PIFA, the IFA can show some Circular Polarization (CP) characteristics. Usually, it is important to place the IFA 2~4 mm away from the PIFA patch because too close a placement would cause the IFA to lose efficiency. On the other hand, if IFA was placed too far away, CP at certain angles could not be achieved. Furthermore, the coupling between the IFA and the PIFA also affects the resonant frequency.

In such a dual-antenna model, another question is how to choose the feed/ground locations for both antennas. For the PIFA, the antenna efficiency and the BW are the driving factors. While for the GPS IFA, besides the efficiency, the polarization has to be taken into account. As is well known, GPS signals from satellites are RHCP. To polarization match a satellite signal in the direction away from the head when a user is holding a handset, RHCP in the antenna front direction (Z direction in Figure 3.31) is necessary. Although IFA itself is a linear antenna, when it excites the currents on the ground plane, the whole structure shows some CP characteristics. Whether it is dominated left-handed or dominated right-handed polarization will depend on the feed/ground location of the GPS antenna. Also, before choosing the feed/ground locations, the distance between the feed and the ground is fixed at 6 mm for the PIFA and 2.5 mm for the IFA. Reasonable BW could be achieved for both antennas with such distances.

It has been demonstrated[32] that in maintaining the Q value ($Q = 1/BW$, where BW is the bandwidth), the antenna efficiency will get higher as the antenna effective size gets larger. For the PIFA, the achievable BW was quite limited, so in order to get a larger effective size, the feed and ground pins had to be put on the top of the patch as shown in Figure 3.31. Only in this way could the ground plane be fully excited to achieve the maximum antenna effective size. For the IFA, since the BW requirement in the GPS band was not as critical, the location of the feed and ground pins was slightly more flexible. The antenna efficiency e_0 is broken down into[33]

$$e_0 = e_r e_c e_d,$$

where e_r is the reflection efficiency, e_c is the conduction efficiency, and e_d is the dielectric efficiency. In the dual-antenna module, the isolation between the two antennas has to be included. Then the efficiency becomes

$$e_0 = e_r e_t e_c e_d = (1 - |S_{11}|^2)(1 - |S_{21}|^2)e_c e_d,$$

where e_t is the isolation efficiency, and S_{11} and S_{21} represent the S-parameters if the dual-antenna model is viewed as a two-port network. The above equation shows that reducing S_{21} increases the total efficiency. Based on the above criteria, antenna model 1 was chosen out of the five antenna model candidates listed in Table 3.3 Antenna model 5 was symmetric to antenna model 1, but this configuration had LHCP in the front direction.

Figure 3.32 shows the measured radiation patterns of antenna models 1 and 5. The source antenna was an RHCP spiral antenna. It is clear that when the feed and ground pins were on the left side, a null appeared in the Z direction in the radiation pattern. If the handset was put against the users head, –Z direction was where the satellite signals came from. This phenomenon can be explained by Figure 3.33. When the feed/ground location is at the right side, because

Table 3.3 Comparison of the PIFA-IFA module with different feed/ground locations. (The efficiency numbers were computed at the center frequency of each band.)

	1	2	3	4	5
S_{21} at 800 MHz band	< -15 dB	< -15 dB	< -15 dB	< -20 dB	< -15 dB
S_{21} at 1900 MHz band	< -10 dB	$-7\sim-8$ dB	< -9 dB	$-7\sim-9$ dB	< -10 dB
S_{21} at GPS band	< -15 dB	$-3\sim-5$ dB	$-4\sim-7$ dB	$-7\sim-8$ dB	< -15 dB
Front polarization of GPS	RHCP	RHCP	RHCP	RHCP	LHCP
Efficiency at 800 MHz band	86%	86%	85%	86%	86%
Efficiency at 1900 MHz band	64%	52%	69%	58%	64%
Efficiency at GPS band	81%	27%	25%	75%	81%

	PIFA patch
——	GPS IFA
(⦂)	GPS feed/ground
●	Feed
○	Ground

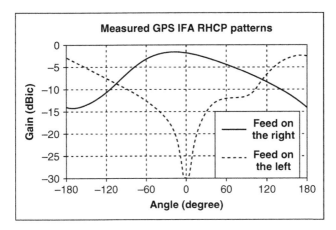

Figure 3.32 Effect of GPS IFA feed/ground location on the polarization; patterns were measured at the XZ plane.

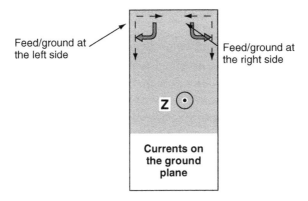

Figure 3.33 Currents excited by the GPS antenna located at the +Z direction on top of the ground plane.

the horizontal length of the ground plane is shorter than the vertical length, the horizontal current will lead in phase. Therefore, dominated right-handed polarization is created. On the other hand, when the feed/ground location is at the left side, dominated left-handed polarization is created.

Similarly, the other antenna models symmetric to models 2 through 4 were eliminated because of the polarization mismatch. Table 3.3 also clearly shows that the isolation between the two antennas directly affected the total efficiency.

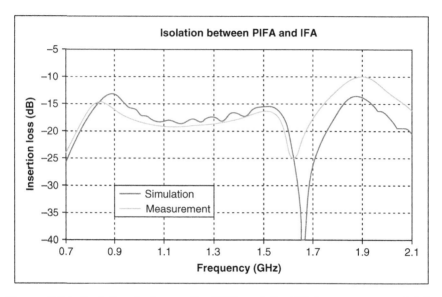

Figure 3.34 Isolation between the PIFA and the IFA in antenna model 1.

A prototype of antenna model 1 was fabricated and measured. The S_{21} was compared to the simulation in Figure 3.34. The plot shows they are quite close except at the 1900 MHz band, where there is more loss due to the coaxial cable in the real measurement. In Figure 3.35, the measured two-dimensional (2-D) radiation patterns of both the PIFA and the IFA all matched the simulation results quite well. The efficiency was not compared because the three-dimensional (3D) efficiency measurement was decided to be too difficult to calibrate in the available chamber for an absolute measurement. A relative efficiency between various IFA lengths was selected since a relative comparison could be done more easily. Another reason that a comparison between measured and simulated data was not done is that the loss in the dielectric and cables was determined to be uncertain for a simulation.

3.3.3.3 Optimization of the IFA Length

Once the locations of the feed and ground legs of both antennas were fixed to further optimize the efficiency of the design, GPS IFAs of three different lengths were tuned to 1575 MHz for comparison. Since the IFA is a narrow-band antenna, with chip inductor-capacitor (LC) components, it was easily tuned to the desired frequency for each length. As the IFA length changed, the coupling between the PIFA

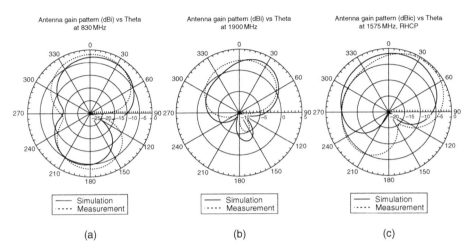

Figure 3.35 Radiation patterns of the PIFA-IFA combination in model 1 at the XZ plane, E_θ. (a) Radiation patterns of the PIFA at 830 MHz; (b) radiation patterns of the PIFA at 1900 MHz; and (c) radiation patterns of the IFA at 1575 MHz.

and the IFA was expected to change and affect the efficiency of both antennas.

Starting from the regular GPS IFA without any matching, the short IFA was made 5 mm shorter, while the long IFA was made 4 mm longer. Both the short IFA and the long IFA were matched to 1575 MHz with LC circuits. The return losses of the PIFA and the IFA are shown in Figures 3.36 and Figure 3.37. The original PIFA with no IFA (referred to as PIFA only) is also shown in Figure 3.36 as the reference antenna.

Clearly, both the PIFAs and the IFAs are matched well. Figure 3.38 shows the isolation curves. Figure 3.37 shows that as the IFA becomes shorter, a second resonance starts to approach the 1900 MHz band. This is also reflected in Figure 3.38, as the isolation becomes worse at about 2.1 GHz for the short IFA case.

On the other band, as the IFA becomes longer, a second resonance starts to approach the 800 MHz band, leading to worse isolation for the long IFA case in the 800 MHz band. A notch is also observed between the GPS band and the 1900 MHz band in Figure 3.38, which moves toward the GPS band as the IFA becomes longer. If designed properly, it is expected that this notch can be moved right to 1575 MHz and the best isolation can be achieved for the GPS band. However, this would be at the price of sacrificing the isolation in the 800 MHz band.

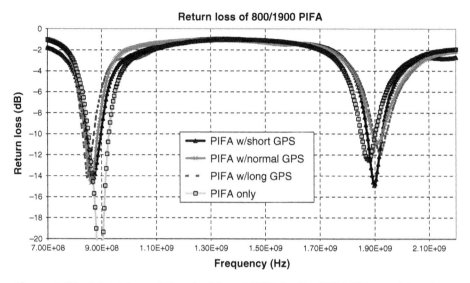

Figure 3.36   Matching of the dual-band PIFA in the PIFA-IFA combination.

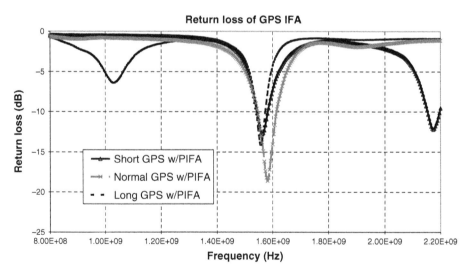

Figure 3.37   Matching of the JFA in the PIFA-IFA combination.

Accordingly, the measured 3D efficiencies in the free space are shown in Figure 3.39. In the 800 MHz band, as the IFA becomes shorter, the isolation and the PIFA efficiency both improve. The efficiency of the PIFA with the short IFA is very close to the efficiency

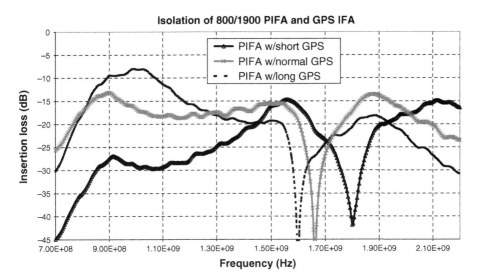

Figure 3.38   Isolation between the IFA and the PIFA in the PIFA-IFA combination.

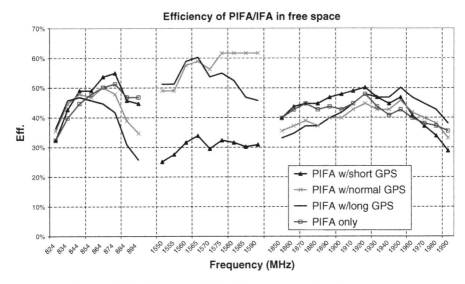

Figure 3.39   Measured 3D efficiencies in the free space.

of the PIFA-only case. In the GPS band, the efficiencies of both the long IFA and the short IFA are lower than the efficiency of the regular IFA. The major reason is the loss caused by the LC matching components. However, the long IFA efficiency is much closer to the regular

IFA because the isolation of the long IFA is much higher (about –26 dB). In the 1900 MHz band, the situation is more complicated. Since the transmitting band (1850~1910 MHz) of the PIFA with the short IFA falls into the notch, the isolation is therefore much higher than the other two cases, resulting in the improved efficiency. While the receiving band (1930~1990 MHz) is impacted by the second resonance as discussed above, the efficiency reduces very quickly. The PIFA with the long IFA shows improved efficiency over the PIFA with the regular IFA because the isolation is higher.

As a summary, depending on the design requirement, to optimize the PIFA at the 80 MHz band, the short IFA was preferred. To optimize the PIFA at the 1900 MHz band, the long IFA should be chosen. In the GPS band, the longer IFA showed improved isolation, but the efficiency was impacted by the loss of the matching circuits. Higher Q LC components with less resistance could have been used for matching to improve the GPS efficiency, but were not available.

3.4   Human Interaction in Handset Antenna Design

In antenna designs for handsets, EM interaction between the human head and the various antennas has been a key factor.[16] On one hand, the human head will affect the antenna performance once the handset is put against the head. Most of the time, antenna impedance match and radiation efficiency will both be degraded. On the other hand, human tissue will be exposed to the EM radiation and cause health concerns in users. In the handset industry, SAR is used as the indication of how much EM radiation the human tissue absorbs. In the United States, the FCC sets the standard to be 1.6 mW/g (1.6 W/kg) averaged over 1 g of tissue. In Europe, the standard is 2.0 mW/g (2.0 W/kg) averaged over 10 g of tissue.

These SAR requirements pose a big challenge on handset antenna designs, which is to design a high efficient antenna in close vicinity to the human body. It seems that this can be achieved by designing an antenna in the handset with a good back-to-front ratio, in which less power will be radiated toward the human in the user position. However, both back-to-front ratio and antenna efficiency are measured in terms of the far field, and it is very difficult to relate the SAR distribution of an exposure to the structure of the incident EM fields. SAR actually represents a complicated near-field interaction between the antenna and the nearby structure. Therefore, it is more important to understand the near-field distribution and how it affects SAR.

In this section, the human effect on the antenna impedance matching and the antenna efficiency will first be studied. Then the method of SAR measurements will be introduced. SAR measurement data of the PIFA-IFA combination will be demonstrated to address potential SAR reduction techniques.

3.4.1   Human Head Effect on Handset Antennas

When antennas are designed for handsets, it is necessary to take into account that handsets are used close to the head of the user, which will absorb a considerable amount of transmit power and therefore will affect the antenna impedance match, radiation patterns, and antenna efficiencies.[16,36,37] Figure 3.40 shows the phantom head used in the against-head antenna measurement. This phantom head is filled with simulated brain fluid.

3.4.1.1   Head Effect on the Impedance Match
As a load to the antenna, the head absorbs a certain amount of EM radiation. However, since the ground plane plays the role of RF shielding, the impedance can be maintained to a certain level. In

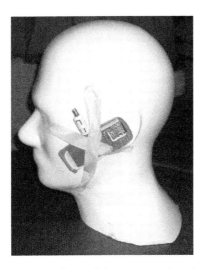

Figure 3.40   The phantom head, used for against-head measurement, is filled with fluid to simulate the brain and other tissues.

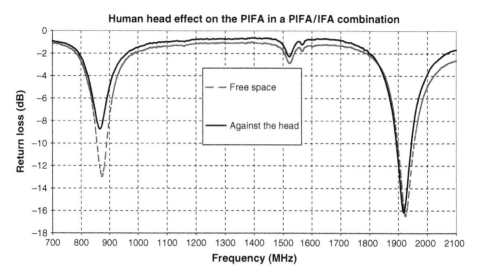

Figure 3.41 Head effect on the impedance match of the PIFA in the PIFA-IFA combination (regular IFA length).

Figures 3.41 and 3.42, the return losses of both the PIFA and the IFA in a PIFA-IFA combination are shown to demonstrate the head effect on the input impedance. At the 800 MHz band, the PIFA is detuned a little, and the BW becomes narrower. At the 1900 MHz band and

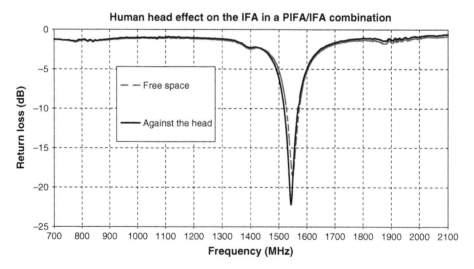

Figure 3.42 Head effect on the impedance match of the GPS IFA in the PIFA-IFA combination (regular IFA length).

the GPS band, the impedances are not affected much by the phantom head.

Other antennas such as the whip antenna or the stubby antenna will show even less effect because those antennas are sticking out of the handsets. Therefore, the head will not block the antenna as much as it does in the PIFA-IFA combination. The hand will also affect the impedance,[16] but in this chapter hand effects are not included.

3.4.1.2   Head Effect on Radiation Patterns

Although not much difference was observed from the return loss measurement between the cases with and without the head, the radiation patterns were found to be significantly changed due to the head, as shown in Figure 3.43. The back radiation (180°) was reduced with the presence of the head, especially at the 800 MHz band. For the case of the PIFA at the 1900 MHz band and the case of GPS IFA at 1575 MHz, the front radiation with the presence of the head maintained a similar level as that in the free space.

In the following Section 3.2.2.1, the comparison between the active whip and the parasitic whip with the presence of the head is shown in Figure 3.44. At the 800 MHz band, the parasitic whip behaves the same as the active whip, while at the 1900 MHz band, the pattern of the active whip still shows high back radiation toward the head. It

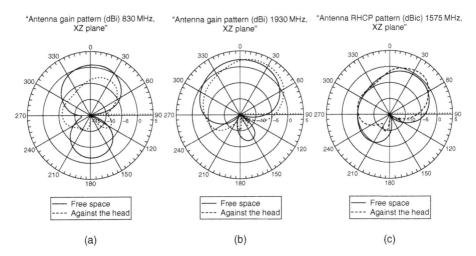

(a)   (b)   (c)

Figure 3.43 Head effect on the radiation patterns of the PIFA in a PIFA-IFA combination. (a) Radiation patterns of the PIFA at 830 MHz; (b) radiation patterns of the PIFA at 1930 MHz; and (c) radiation patterns of the IFA at 1575 MHz.

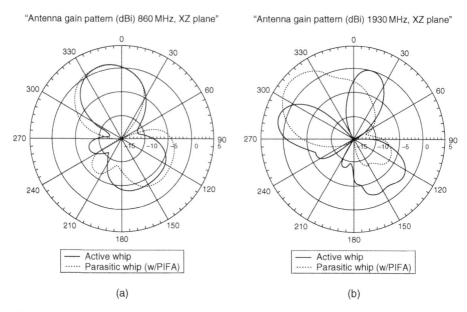

Figure 3.44 Head effect on the radiation patterns of whip antennas, and comparison of the active whip and the parasitic whip.

is also observed by comparing Figures 3.43(a) and 3.44(a), with the presence of the head, that the whip improved the gain of the PIFA by up to 3 dB at the 800 MHz band, which is the major reason that the whip still plays an important role in current handset antenna designs.

3.4.1.3 Head Effect on Antenna Efficiencies

It is expected that with the handset put besides the head, a certain amount of radiated power will be absorbed by the head. Comparing Figures 3.39 and Figure 3.45, antenna efficiencies of the PIFA dropped about 20~25% at the 800 MHz band and 15~20% at the 1900 MHz band. Antenna efficiencies of the IFA dropped about 15~30%.

Although the reduction of the antenna efficiency caused by the head cannot be avoided, some method can be applied to improve the efficiency and minimize the reduction due to the head. As shown in Figure 3.45, at the 1900 MHz band, the efficiencies of all three cases with the IFA at the against-head position are better than that of the PIFA-only case. This indicates that the IFA actually acts as a directive element, which improves the PIFA's efficiency and can possibly be used for the SAR reduction.

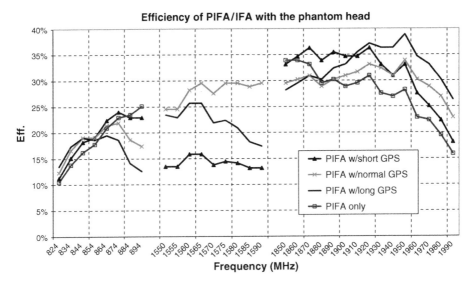

Figure 3.45   Measured 3D efficiencies of the PIFA-IFA combination against the phantom head.

3.4.2   SAR Consideration in Handset Antenna Designs

The EM radiation toward the head has raised a common public concern on health. Up to now, there has been no clear scientific evidence that the radiation from the handset is hazardous to human beings. However, to quantify the human exposure to RF energy, SAR has been used.[16]

3.4.2.1   Definition of SAR

In the near field of RF sources, the EM fields are very complicated, and the normal incident power density cannot be defined. Since human tissue is composed mainly of water, electrolytes, and complex molecules, it extracts energy from the RF E field by ionic motion and oscillation of polar molecules.[11] The total power absorbed in the tissue is defined as

$$P_{abs} = \frac{1}{2} \int_{v} \sigma |E|^2 dV, \tag{3.6}$$

and SAR is defined as

$$SAR = \frac{\sigma}{2\rho} |E|^2, \tag{3.7}$$

where σ is the conductivity of the medium (usually human tissue), ρ is the mass density, and E is the peak electric field. SAR is expressed in units of watts per kilogram (W/kg) or milliwatts per gram (mW/g).

Clearly, based on Chapter 2, Eq. (2.7), SAR is a point quantity, and its value varies from one location to another. The accuracy and reliability of a given SAR value would depend on three parameters: tissue density, conductivity, and electric field. However, the most significant one of the three parameters is the induced electric field. In the antenna design, the target is to reduce the electric field that is induced in the head.

The induced electric field is a complex function of frequency, handset size, distance between the handset and the head, polarization, and biological variables such as the tissue type. The typical ingredients of the head model are listed in Table 3.4.[38]

SAR can be obtained from detailed numerical computations.[39] Many different tools have been developed to simulate SAR. Finite-Difference Time-Domain (FDTD)[40] is quite popular for SAR calculation because this algorithm allows antennas to be modeled in their true operating environment.[16] SemCAD is the commercial tool which is integrated into the SPEAG DASY 4 SAR testing system. It also uses FDTD to calculate SAR. The Finite Element Method (FEM)[41] is another popular method for SAR simulation and is used in Ansoft HFSS, which has integrated SAR calculation macros for convenient computation. Depending on the human head model and the antenna model generated by the simulation tool, SAR values calculated could be quite accurate. When it comes to a real handset, it is still difficult to simulate

Table 3.4   Values of tissue parameters (relative permittivity ε_r effective conductivity σ, density ρ)

Tissue	900 MHz		1800 MHz		
	ε_r	σ (S/m)	ε_r	σ (S/m)	ρ (kg/m³)
Cartilage	42.65	0.782	40.21	1.287	1100
Muscle	55.95	0.969	54.44	1.389	1050
Eye	55.27	1.167	53.57	1.602	1020
Brain	45.8	0.766	43.54	1.153	1040
Dry skin	41.4	0.867	38.87	1.184	1090
Skull (bone)	16.62	0.242	15.56	0.432	1645

all the variables. So the numerical modeling of a handset for SAR can be costly, and it can take as long as several weeks due to the large amount of computations. Consequently, accurate SAR values are still very difficult to achieve without the measurement.

3.4.2.2   SAR measurement

SAR can be measured directly using body/head phantoms, robot arms, and associated test equipment. The DASY 4 robotic system as shown in Figure 3.5 is used to measure the actual SAR. By using conventional SAR measurements, one single test (one position and one frequency) takes about 15 minutes. To complete the head SAR measurement of one dual-band handset, it will take roughly 1 week. Given the large amount of reflection and scattering that a human body causes in the near field, it is typically necessary to measure the RF power deposited inside the tissue rather than the external incident EM fields. In the DASY system (FCC compliant for conducting SAR tests), the SAR phantom is filled with simulated human tissue fluid. The electric parameters of the fluid are listed in Table 3.6 The robot-controlled electric probe is programmed to measure the electric field in volts per meter (V/m) inside the SAR fluid during the operation. The handset under test is required to be performed at the maximum output power, which is intended to represent the worst-case scenario.

The handset is usually measured at two positions in the head SAR measurement, namely, the cheek position and the tilt position, as shown in Figure 3.46. The cheek position is to put the handset against the head with both the ear and the cheek touching the handset.

Table 3.5   Ingredients of the test fluid for the FCC head SAR measurement.

800 MHz band		1900 MHz band	
Ingredient	Head (% by weight)	Ingredient	Head (% by weight)
Deionized water	51.07	Deionized water	54.88
HEC	0.23	Butyl diglycol	44.91
Sugar	47.31	Salt	0.21
Preservative	0.24		
Salt	1.15		

Table 3.6 Electric parameters of the head tissue stimulant.

| | | Head tissue simulant | | |
| | | Dielectric parameters | | |
f(MHz)	Description	ε_r	σ (S/m)	Temp (°C)
836.5	Recommended value	41.5	0.9	21
	+/–5% window	39.4–43.6	0.86–0.95	N/A
1880	Recommended value	40	1.4	21
	+/–5% window	38–42	1.33–1.47	N/A

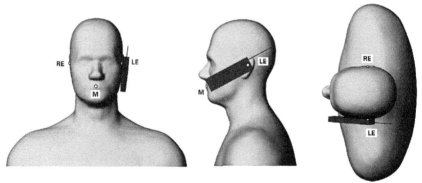

(a) "Touching cheek" position. The phone is angled from mouth to ear with the center of the phone's speaker aligned to the ear reference point (LE or RE)

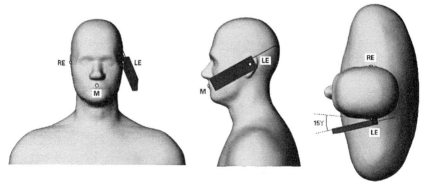

(b) "15° tilt" position. The phone remains on the mouth-to-ear angle (as for the "touching cheek" position), but the microphone end is rotated away from the head by 15°.

Figure 3.46 SAR test positions.

The tilt position is to rotate the handset 15° from the cheek position. These two positions represent the two most typical user positions. Since the head can have an impact on the impedance of the antenna, the actual curve of the handset front cover can be carefully designed to give the optimum SAR performance. However, antenna designers usually do not get involved in the mechanical design of the handset. As a result, some other methods have to be applied for SAR reduction.

During the SAR measurement, the area scan (user-defined area, also called coarse scan) is first conducted to determine the field distribution, as shown in Figure 3.47(a). Then the zoom scan is performed around the highest E-value spot (hot spot) to determine the average SAR value, as shown in Figure 3.47(b). The maximum SAR value is averaged over a cube of tissue using interpolation and extrapolation. The interpolation of the points is done with 3D-Spline, while the extrapolation is based on a least square algorithm.[42] If the hot spot is at the border of the area scan, an enlarged area will be defined and the area scan will be repeated.

The measurement uncertainties are defined in the IEEE P1528 specification. Overall uncertainties must be below 30% for a 95% confidence level. An uncertainty in measurements of 30% may seem a little high, but this percentage is small in decibel terms.

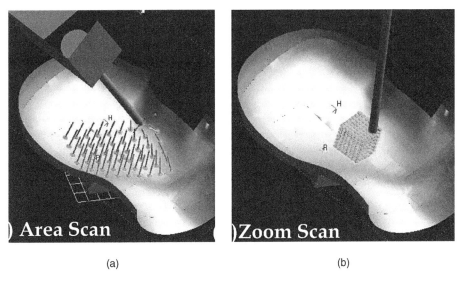

(a)   (b)

Figure 3.47   Scan procedures in the SAR measurement.

3.4.3   SAR Reduction with a GPS IFA

One of the targets for the antenna designs is to find an efficient method to lower the SAR values while minimizing the effect on antenna efficiencies. A lot of research has been conducted in this area[16,38,43,44] in which both SAR and the antenna efficiency were thoroughly investigated. In Jensen and Rahmat-Samii,[16] different types of PIFAs were compared, while Vaughan and Scott[43] focused more on whip antennas. In Kirekas et al.,[38] attention was paid to the effect that the handset chassis had on SAR and antenna efficiencies. In addition to these studies, various SAR reduction techniques were introduced in References[45,50]. Increasing the distance between the handset and the head was a typical method where an earpiece was attached to the handset.[45,46] Using RF absorbing shielding materials[45,47] was another effective way, but this required careful selection of the materials and could potentially reduce the antenna efficiency. Parasitic metal elements have been used to alter the antenna near-field distribution[48] and lower the SAR value of handset[49,50] with some limited success.

In Section 3.4.1.3, it was shown that, at the 1900 MHz band and at the against-head position, the efficiency of the cellular PIFA with different lengths of GPS IFA is better than that of the PIFA-only case. This indicates that the GPS IFA actually improves the PIFA's efficiency at the 1900 MHz band when the antenna is put against the head. At the 800 MHz band, there is no such impact from the GPS IFA. How will this affect the SAR value?

Table 3.7 shows the relative SAR measurement results at the cheek position. SAR at the tilt position was also measured, but the values are lower for all the cases than SAR values at the cheek position. During the SAR measurement, the whole antenna model was put into a plastic housing to simulate the handset chassis, which provides 8 mm of

Table 3.7 SAR measurements of the PIFA in the PIFA-IFA combination. (The data is referenced to the measurement of the PIFA-only case, either as a percentage increase or decrease.)

Frequency (MHz)	824	836	849		1850	1880	1910
PIFA w/short IFA	0.52%	0.85%	−2.17%		12.6%	22.6%	49.5%
PIFA w/regular IFA	2.78%	1.09%	−3.74%		−30.7%	−31.2%	−23.9%
PIFA w/long IFA	−0.52%	1.91%	−2.17%		−30.7%	−31.1%	−29.8%

separation between the antenna ground plane and the earpiece of the SAR phantom head. A prototype transmitter unit was also integrated to eliminate the coaxial cable. Also, the efficiency difference was taken into account to adjust the transmitting power. For example, if the efficiency at a certain frequency was 0.5 dB lower than the reference, then a 0.5 dB higher transmitting power was used for the SAR measurement. As in reality, SAR was only measured at the transmitting band (824~849 MHz at 800 MHz band, 1850~1910 MHz at 1900 MHz band). Measurements of 824, 836, and 849 MHz were chosen to represent the low, middle, and high frequencies in the 800 MHz transmitting band. At the 1900 MHz band, 1850, 1880, and 1910 MHz were chosen.

As expected, for all the cases, little SAR difference was observed at the 800 MHz band in Table 3.7. Figure 3.48 shows hot spots (the square area, where the field is concentrated) and SAR distributions by the area scan (user-defined area for field strength scan to find the hot spot). In Figure 3.48, it is the bigger square area that covers the handset with the grid inside. Actual SAR values were obtained by the zoom scan in the hot spot. It is obvious that all the cases are very similar.

However, at the 1900 MHz band, one observes surprisingly different values in Table 3.7. Compared to the PIFA-only case, the PIFA with the short IFA shows an average of 20% SAR increase. On the other hand, for the cases of the long and regular IFAs, SAR values are 30% lower. This indicates that, with enough length, the IFA performed the role of the parasitic element to alter the near-field distribution and to provide SAR reduction. Two hot spots are observed in the SAR distribution in Figure 3.49. The dark dots inside the hot spots were automatically located by the measurement system as long as the field strength of the

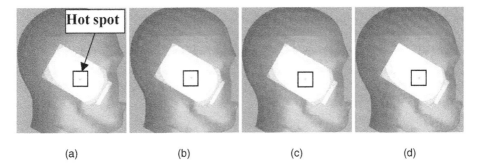

(a)   (b)   (c)   (d)

Figure 3.48 Area scan of the SAR measurement at the 800 MHz band. (a) PIFA only; (b) PIFA w/short IFA; (c) PIFA w/regular IFA; and (d) PIFA-IFA w/long IFA.

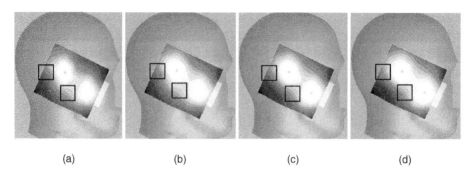

(a)   (b)   (c)   (d)

Figure 3.49   Area scan of the SAR measurement at the 1900 MHz band. (a) PIFA only; (b) PIFA w/short IFA; PIFA w/regular IFA; and (d) PIFA w/long IFA.

second hot spot was within 2 dB differences from the first hot spot. In Figures 3.49(a) and (b), the hot spot near the ear is much stronger than the other hot spot down to the cheek, which indicates that all the field energy is concentrated near the ear hot spot. In the cases of Figures 3.49(c) and (d), the energy is spread up between the two hot spots, which results in the lower SAR.

It has been demonstrated that in the PIFA-IFA combination, the GPS IFA can be viewed as a director for the PIFA. It will change the near-field interaction between the antenna module and the head and therefore will achieve SAR reduction by spreading the energy between the SAR hot spots. In the next chapter, another SAR reduction technique will be introduced, in which the key is also to balance the SAR hot spots.

3.5   Total Radiated Power (TRP)

In the wireless industry, TRP is a measurement that correlates well to the field performance of the handset. As a result, the CTIA has specified TRP as one of the over-the-air performance tests for handsets.[18]

TRP is influenced by both the TX power, which must be low enough to meet the FCC SAR requirements, and the antenna efficiency. The TX power in a handset will typically be determined by one of two limitations. One limitation is how much power the PA can deliver to the antenna, while maintaining sufficient linearity to maintain signal integrity. The other limitation is that since SAR is directly proportional to the TX power, then the TX power of a handset cannot exceed the level, which corresponds to exceeding the FCC SAR limit. In many cases, it is the SAR that limits the TX power rather than the maximum

PA output power. It has been found that handsets with higher efficiency antennas might not necessarily have higher TRP. Our results consider SAR and TRP at the same time.

This section will follow the previous section on SAR reduction techniques. The effect that different grounding methods have on SAR will be discussed by comparing four types of PIFAs that were designed for the 1900 MHz band. The ground plane, the dielectric substrate, and the feed were fixed for all the antenna models. It is observed that the ground connection does not only cause the radiation patterns to change, but also cause the energy to spread between the two SAR hot spots and, consequently, change the SAR. SAR values of all the antenna models were normalized to the same level by adjusting the TX power. The TRPs were then calculated.

3.5.1   Definition of TRP

Very often, high antenna efficiency and low SAR in a handset cannot be achieved at the same time. Then how can one evaluate the antenna designs in handsets? TRP provides a good evaluation of the antenna performance from the system point of view. It combines the antenna efficiency and the TX power limited by SAR together.

The radiated power is defined as

$$P_{rad} = \frac{1}{2}\text{Re}\left\{\int_S E \times H^* \cdot \hat{n} dS\right\}, \tag{3.8}$$

where E and H are the electric and magnetic field intensities on a surface S completely enclosing the antenna structure. For an active handset (without cable), the TRP is equal to the measured P_{rad}. However, directly measuring TRP is very difficult due to the expensive equipment involved in the active test setup.

From Chapter 2, Equation 2.5, we get

$$P_{rad} = e_0 \cdot P_{del}. \tag{3.9}$$

By transferring this result in decibel units and applying it to an active handset, TRP can be achieved by

$$TRP = transmit\ power + antenna\ efficiency, \tag{3.10}$$

where TX power is equivalent to P_{del}, which is limited by the SAR requirement and is usually converted from mW to dBm; the antenna efficiency (e_0) can also be converted from percentage to dB. In this way, the antenna efficiency can be obtained with the passive measurement using coaxial cables, while the TX power can be obtained during the SAR measurement.

3.5.2   PIFA Models in the 1900 MHz Band

The configurations of the PIFA models are shown in Figure 3.50. The ground plane size was 98×34 mm. The substrate was $26 \times 34 \times 10$ mm and was made of Polycarbonate Acrylonitrile Butadiene Styrene (PC/ABS), one of the most popular engineering plastics. The dielectric constant (ε') of PC/ABS is about 2.6, and the loss tangent (δ) is about 0.04. Based on the above constraints, four different PIFAs were each tuned to the 1900 MHz band for comparison.

PIFA #1 was a typical single-band design, where the length of the patch was approximately a quarter wavelength. The patch size of PIFA #2 was almost twice that of PIFA #1. In order to achieve the same frequency band tuning, the ground pin was moved to the bottom of the patch. PIFA #3 had the maximum patch size among the first three designs. In this case, only moving the ground pin was not enough to shift the resonant frequency up to the 1900 MHz band. So a 12 mm wide metal strip was used to connect the patch and the ground plane. PIFA #4 had the same size patch as PIFA #3. A simple and efficient J-shape slot was used to achieve a dual-band PIFA to cover both the

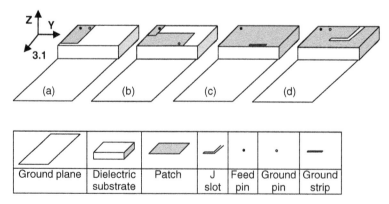

Figure 3.50 Configuration of PIFA models. (a) PIFA #1, patch size 19×9 mm; (b) PIFA #2, patch size 18×20 mm, ground pin moved; (c) PIFA #3, patch size 26×34 mm, 12 mm wide ground strip; and (d) PIFA #4, patch size 26×34 mm, with J slot.

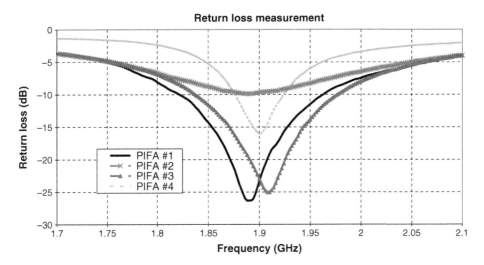

Figure 3.51 Impedance matching of PIFA models.

800 and the 1900 MHz bands. But in this section, only the 1900 MHz band was of interest.

Figure 3.51 shows the return loss measurements of all four designs. Using -6 dB S_{11} as the criteria for the BW, it was expected that PIFA #4 would have the narrowest BW due to the dual-band design. The other three designs had similar BWs. However, PIFA #2 was not matched as well as PIFAs #1 and #3.

Measured antenna efficiencies are shown in Figure 3.52 for the case in the free space and in Figure 3.53 for the case at the against-the-head position. PIFA #1 had the highest antenna efficiencies in both cases. In the free space, the antenna efficiency of PIFA #2 was almost 10% lower than that of PIFA #1, but only about 5% lower with the presence of the phantom head. It is also interesting to observe that the antenna efficiency of PIFA #3 was affected the most by the phantom head. For PIFA #4 at the transmit band (1850~1910 MHz), the antenna efficiencies were comparable to PIFAs #2 and #3, but because of the BW limitation, the antenna in the receive band (1930~1990 MHz) had a lower efficiency.

The radiation patterns in the E plane (defined as the XZ plane in Figure 3.4 are shown in Figure 3.54(a) for the case in the free space and in Figure 3.54(b) for the case at the against-the-head position. For PIFAs #2 and #3, the antenna gain lobes toward the head direction (back lobe) are higher than that of the other two designs, and this is because the positions of the ground pin and the ground strip were

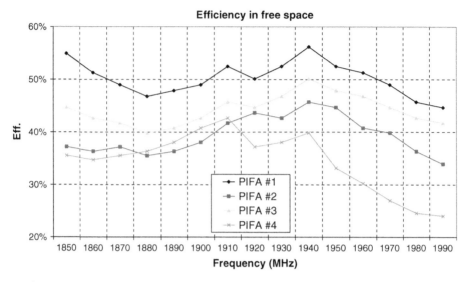

Figure 3.52 Efficiency measurements of PIFA models in the free space.

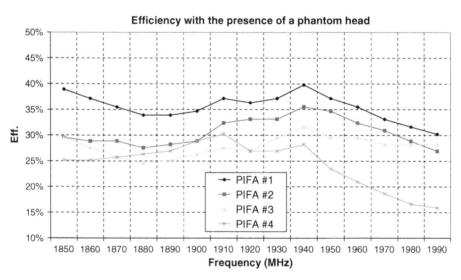

Figure 3.53 Efficiency measurements of PIFA models in the against-the-head position.

located at the bottom of the patch. Notably, the back lobe of PIFA #3 in the free space was the highest, so the antenna efficiency had the greatest reduction when the antenna was placed next to the phantom head.

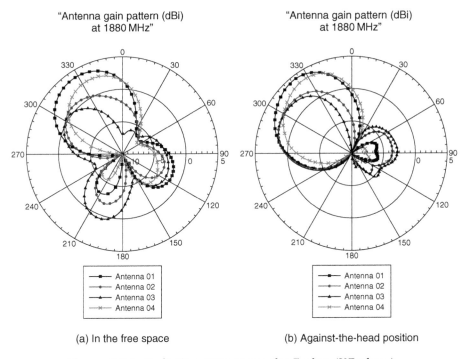

"Antenna gain pattern (dBi) at 1880 MHz"   "Antenna gain pattern (dBi) at 1880 MHz"

(a) In the free space   (b) Against-the-head position

Figure 3.54   Radiation patterns at the E plan (XZ plane).

In terms of the impedance matching and the efficiency results, it was clearly demonstrated that PIFA #1 was the best design. However, when SAR and TRP are both taken into account, one needs to assess if PIFA #1 still exceeds the other three designs.

3.5.3   SAR and TRP

Unlike the antenna efficiency, which is associated with the far field, the SAR is dependent on the near-field interaction between the antenna and the head. Similar to the case of PIFA-IFA study, during the SAR measurement, the whole antenna model was put into a plastic housing to simulate the handset chassis, which provided 8 mm of separation between the antenna ground plane and the ear of the SAR phantom head. A prototype transmitter was also integrated to eliminate the coaxial cable. As with actual handsets, the SAR was only measured at the TX band, so 1850, 1880, and 1910 MHz were picked for the measurement to represent the low, mid, and high frequencies.

SAR was measured with each of the antenna models using the same TX power. However, in this work, the SAR impact on TRP will be addressed by adjusting the TX power level to achieve the same SAR value for all of the cases. So by normalizing the SAR values to 1.0 mW/g for all the PIFA models, the TX power table, shown in Table 3.8, was obtained. Now due to the normalized SAR, PIFA #1 had the lowest TX power level, which also meant that it had the highest SAR value among the four cases when a constant TX power was used. PIFA #2, on the other hand, could accept the highest power delivered from the PA in the handset system, while maintaining the same SAR as the other cases.

The hot spots in the measured SAR distributions illustrated why PIFAs #2 and #3 had better SAR performance. At the cheek position in Figure 3.55, in all cases, two hot spots were observed in the SAR distributions by the area scan. One hot spot was located at the ear, while the other was next to the cheek. The solid square represents the dominated hot spot where the final SAR values were obtained. For PIFA #1 in Figure 3.55(a), the field concentrated near the ear hot spot was very strong and resulted in a high SAR. For PIFAs #2 and #3, since the patch was grounded at the bottom, more energy spread to the hot spot next to the cheek, which consequently reduced the SAR. The lower antenna efficiencies in these cases also contributed to the lower

Table 3.8 Power level comparison of different PIFA models. (SAR was normalized at 1.0 mW/g.)

	Frequency (MHz)	PIFA #1	PIFA #2	PIFA #3	PIFA #4
Transmit power (dBm)	1850	22.1	24.1	23.6	23.4
	1880	22.25	23.85	23.4	22.6
	1910	21.8	23.3	22.9	22.6
TRP in the free space (dBm)	1850	19.5	19.8	20.1	18.9
	1880	18.95	19.35	19.4	18.2
	1910	19	19.5	19.5	18.9
TRP with phantom head (dBm)	1850	18	18.8	18.3	17.4
	1880	17.55	18.25	17.3	16.8
	1910	17.5	18.4	17.3	17.4

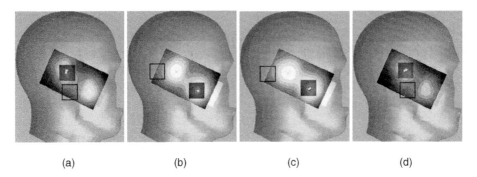

(a)   (b)   (c)   (d)

Figure 3.55   Area scan of the SAR measurement at the cheek position: (a) PIFA #1, (b) PIFA #2, (c) PIFA #3, and (d) PIFA #4.

SAR. Similar to PIFA #1, the hot spot at the ear also dominated for the case of PIFA #4. SAR was lower mainly because the antenna efficiency was lower.

In Figure 3.56, as the handset was tilted, most of the energy was distributed in the ear hot spot. However, it could still be observed that more energy had spread to the cheek for the cases of PIFAs #2 and #3. Back to Figure 3.55, the ground strip in PIFA #3 caused enough energy to be shifted down to the cheek to make it the hottest spot and to result in an increased SAR. It could be projected that the optimum design would be achieved by choosing an optimum width of the ground strip for a certain patch size, which would balance the energy between the two hot spots and, consequently, achieve the lowest SAR for a fixed TX power.

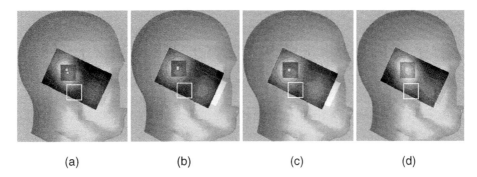

(a)   (b)   (c)   (d)

Figure 3.56   Area scan of the SAR measurement at the tilt position: (a) PIFA #1, (b) PIFA #2, (c) PIFA #3, and (d) PIFA #4.

Although the efficiency of PIFA #1 was the highest among the four PIFA models, SAR was also the highest. So how could one determine which of the four designs was better?

Table 3.8 lists the TRP numbers of all the PIFA models. In the free space, the TRP of PIFAs #2 and #3 is similar and higher than that of PIFAs #1 and #4. With the phantom head, the TRP of PIFA #3 decreased quickly and became slightly lower than the TRP of PIFA #1. Overall, PIFA #2 showed the highest TRP. In the handset system, lowering the TX power improves the battery life and helps to reduce the heat generated by the PA. However, higher TRP can significantly improve the call performance of the handset in a weak signal area. Consequently, the system design must take trade-offs when trying to optimize the TRP, TX power, battery life, and in-use handset temperature.

3.6   Conclusion

The main objective of this chapter is to provide a design basis for the antenna designer working on wireless applications. Validated by the measurement results, two powerful numerical tools, FEM (Finite Element Method) and FDTD (Finite-Difference Time-Domain), are both capable of simulating the antenna structures applicable in a handset. The impedance matching, antenna efficiency, antennas coupling, and human effect on the handset antenna designs have all been covered.

PIFAs are widely used in the handset industry. A number of methods to improve the bandwidth of a dual-band PIFA under certain design restrictions were verified and would be very useful in designing handset antennas. When the ground plane length, patch size, and the substrate dimension were fixed, it was shown that the 1900 MHz band had more flexibility than the 800 MHz band in terms of the bandwidth improvement, because the available patch area vs. the wavelength at the 1900 MHz band was much larger. Among all the methods, modifying the feed structure was the most direct and efficient way. Furthermore, during the industrial design process, it is quite challenging to maintain the dual-band functionality and improve the bandwidth at the same time, since many factors could affect the bandwidth. Most often, trade-offs between the two bands have to be made to achieve the final design goal.

Another common challenge in the industry is to optimally integrate multiple antennas in a handset. The coupling between the antennas can degrade the antenna efficiency. Solutions to minimize the coupling between different antennas were discussed. In the PIFA-Whip combination, the retracted whip causes destructive coupling to the PIFA in a handset. An LC circuit model was developed and verified in RF ADS to simulate the decoupling effect and to assist the HFSS full wave simulation. The simulation time of RF ADS is much shorter than that of HFSS, which results in significantly reduced designing time. By using RF ADS simulation to iteratively evaluate LC variations, values for the LC components can be found which optimally decouple the retracted whip from the PIFA very well.

In another case, a full wave analysis of different antenna models of the PIFA-IFA combination was provided. In this combination, the efficiencies of both antennas were affected by the isolation between them. To optimize the design, the feed/ground pins of both antennas must be located carefully first to achieve better efficiencies. At the same time, matched polarization for the GPS IFA must also be taken into account. By tuning the IFA length, one could optimize the isolation between the two antennas at the particular frequency band depending on the design requirement.

The human head has a strong impact on the antenna performance of the handset. The input impedance, antenna efficiencies, and radiation patterns are all affected. First, the bandwidth usually decreases, especially at the 800 MHz band. Second, the back radiation of the antennas is significantly reduced due to the presence of the head. For a PIFA, the front radiation will even drop $3 \sim 5$ dB at the 800 MHz band, but can maintain the same level at the 1900 MHz band and the GPS band. In addition, two whip solutions were also presented, in which the parasitic whip was degraded by the head more than the active whip due to the degradation of the PIFA.

SAR is another key issue in handset antenna design when including the head effect. In order to understand the RF radiation from the antenna toward the head, SAR is carefully studied in this chapter. The effect on the SAR value due to the IFA length was discussed. It was concluded that as long as the IFA has a adequate length, it could lower the SAR by performing in a positive way to alter the near-field distribution of the PIFA at the 1900 MHz band. This part also indicates one of the SAR reduction techniques in the PIFA design, that by using an IFA as the parasitic director, the RF energy could be spread between two SAR hot spots in the cheek area and therefore SAR could be lowered.

Another perspective when analyzing SAR has also been provided, which is to consider the impact of SAR on TRP. The antenna design affected both the antenna efficiency and SAR, in which increased efficiency correlated closely with an increased SAR. The TX power is usually limited by the SAR requirement. The TRP metric takes into account both SAR and the antenna efficiency, and it correlates well with the handset field performance. As a consequence, antennas with good efficiencies may not necessarily have good TRPs. Four types of PIFAs that were designed for the 1900 MHz band were compared. Among the four PIFA designs, PIFA #1 had the highest efficiency, but also the highest SAR. PIFA #1 did not have the highest TRP. Although the efficiency of PIFA #2 was lower, when including the SAR effect, it showed the highest TRP. This study also leads to another SAR reduction technique for the PIFA design at the 1900 MHz band, which is that by grounding the patch at the bottom, the energy could be balanced between the two SAR hot spots and optimum SAR could then be achieved.

We strongly believe that ongoing research efforts are essential to design novel multi-function antennas for handset units with ever-increasing demanding performance requirements. This must be performed with utmost attention given to the safety issues for which we do not have a clear understanding yet.

References

1. FCC Report and Order, Docket No. 94-102, "Revision of the Commission Rules to Ensure Compatibility with Enhanced 911 Emergency Calling Systems," RM-8143, adopted in June 1996.
2. "Wireless Location Services: 1997," Strategies Group, 1997.
3. J. J. Caffery and G. L. Stuber, "Overview of Radiolocation in CDMA Cellular Systems," *IEEE Commun. Mag.*, pp. 38–45, April 1998.
4. FCC Report and Order, Docket No. 01-330, "Revision of the Commission Rules Affecting the Conversion to Digital Television," adopted in November 2001.
5. J. R. Tuttle, "Traditional and Emerging Technologies and Applications in the Radio Frequency Identification (RFID) Industry," *IEEE Radio Frequency Integrated Circuits (RFIC) Symposium*, pp. 5–8, June 1997.
6. R. Bansal, "Coming Soon to a Wal-Mart Near You," *IEEE Antennas Propagat. Mag.*, Vol. 45, pp. 105–106, December 2003.
7. J. H. Winters, J. Salz, and R. D. Gitlin, "The Impact of Antenna Diversity on the Capacity of Wireless Communication Systems," *IEEE Trans. Commun.*, Vol. 42, pp. 1740–1751, April 1994.

8. W. C. Y. Lee, *Mobile Communications Engineering*. New York: John Wiley & Sons, 1982.

9. M. A. Jensen and Y. Rahmat-Samii, "Performance Analysis of Antennas for Hand-held Transceivers Using FDTD," *IEEE Trans. Antennas Propagat.*, Vol. 42, pp. 1106–1113, August 1994.

10. K. L. Virga and Y. Rahmat-Samii, "Low-Profile Enhanced-Bandwidth PIFA Antennas for Wireless Communications Packaging," *IEEE Trans. Antennas Propagat.*, Vol. 45, pp. 1879–1888, October 1997.

11. K. Fujimoto and J. R. James, Editors, *Mobile Antenna Systems Handbook, Second Edition*. Norwood, MA: Artech House, 2001.

12. T. Taga, "Performance Analysis of a Build-In Planar Inverted F Antenna for 800 MHz Band Portable Radio Units," *IEEE Selected Areas Commun.*, Vol. SAC-5, pp. 921–929, June 1987.

13. K. L. Wong, *Planar Antennas For Wireless Communications*. New York: John Wiley & Sons, 2003.

14. S. Eggleston and S. Lahti, "Antenna Transducer Assembly, and an Associated Method Therefore," U.S. Patent No. 6618011, September 9, 2003.

15. R. Kronberger, H. Lindenmeier, L. Reiter, and J. Hopf, "Multiband Planar Inverted-F Car Antenna for Mobile Phone and GPS," *IEEE International Symposium on Antennas & Propagation*, Orlando, FL. July 1999.

16. M. A. Jensen and Y. Rahmat-Samii, "EM Interaction of Handset Antennas and a Human in Personal Communications," *Proc. IEEE*, Vol. 83, No. 1, pp. 7–17, January 1995.

17. J. S. Colburn and Y. Rahmat-Samii, "Human Proximity Effects on Circular Polarized Handset Antennas in Personal Satellite Communications," *IEEE Trans. Antennas Propagat.*, Vol. 46, pp. 813–820, June 1998.

18. Cellular Telecommunications & Internet Association Method of Measurement for Radiated RF Power and Receiver Performance Test Plan, Revision 2.0, March 2003.

19. N. Chavannes, R. Tay, N. Nikoloski, and N. Kuster, "Suitability of FDTD-Based TCAD Tools for RF Design of Mobile Phones," *IEEE Antennas Propagat. Mag.*, Vol. 45, pp. 52–64, December 2003.

20. D. Heberling, "Modern Trends in the Development of Small and Handy Antennas," *Proc. IEEE MTT-S*, Vol. 1, pp. 475–480, July 2001.

21. W. L. Stutzman and G. A. Thiele, *Antenna Theory and Design*. New York: John Wiley & Sons, 1998.

22. R. C. Johnson and H. Jasik, *Antenna Engineering Handbook*. New York: McGraw-Hill, 1984.

23. K. L. Wong, G. Y. Lee, and T. W. Chiou, "A Low-Profile Planar Monopole Antenna for Multiband Operation of Mobile Handsets," *IEEE Trans. Antennas Propagat.*, Vol. 51, pp. 121–125, January 2003.

24. J. M. Zagami, S. A. Parl, J. J. Bussgang, and K. D. Melillo, "Providing Universal Location Services Using a Wireless E911 Location Network," *IEEE Commun. Mag.*, Vol. 36, pp. 66–71, April 1998.

25. M. Ali, R. A. Sadler, and G. J. Hayes, "A Uniquely Packaged Internal Inverted-F Antenna for Bluetooth or Wireless LAN Application," *IEEE Antennas Wireless Propagat. Lett.*, Vol. 1, pp. 5–7, 2002.

26. M. Hirvonen, P. Pursula, K. Jaakkola, and K. Laukkanen, "Planar Inverted-F Antenna for Radio Frequency Identification," *Electronics Lett.*, Vol. 40, pp. 705–707, July 2003.

27. P. R. Foster and R. A. Burberry, "Antenna Problems in RFID Systems," *IEE Colloquium, RFID Technology (Ref. No. 1999/123)*, pp. 3/1–3/5, October 1999.

28. R. W. Waugh, "Handset PA-Duplexer Interaction When the Isolator Is Eliminated," *High Frequency Electronics*, pp. 30–34, March 2004.

29. Z. Li and Y. Rahmat-Samii, "Whip-PIFA Combination in Wireless Handset Application: A Hybrid Circuit Model and Full Wave Analysis," *IEEE International Symposium on Antennas & Propagation*, Monterey, CA. June 2004.

30. M. C. Huynh and W. Stutzman, "Ground Plane Effects on Planar Inverted-F Antenna (PIFA) Performance," *IEE Proc. Microwave Antenna Propagat.*, Vol. 150, pp. 209–213, August 2003.

31. A. T. Arkko and E. A. Lehtola, "Simulated Impedance Bandwidth, Gains Radiation Patterns and SAR Values of a Helical and a PIFA Antenna on Top of Different Ground Planes," in *IEE 11th International Conference on Antennas and Propagat.*, pp. 651–654, April 2001.

32. R. C. Hansen, "Fundamental Limitations in Antenna," *Proc. IEEE*, Vol. 69, No. 2, pp. 170–182, February 1981.

33. C. A. Balanis, *Antenna Theory Analysis and Design.* New York: John Wiley & Sons, 1997.

34. K. Boyle, "Radiation Patterns and Correlation of Closely Spaced Linear Antennas," *IEEE Trans. Antennas Propagat.*, Vol. 50, pp. 1162–1165, August 2002.

35. S. Blanch, J. Romeu, and I. Corbella, "Exact Representation of Antenna System Diversity Performance from Input Parameter Description," *Electronics Lett.*, Vol. 39, pp. 705–707, May 2003.

36. V. Hombach, K. Meier, M. Burkhardt, E. Kuhn, and N. Kuster, "The Dependence of EM Energy Absorption upon Human Head Modeling at 900 MHz," *IEEE Trans. Microwave Theory Tech.*, Vol. 44, pp. 1865–1873, October 1996.

37. K. Meier, R. Kastle, V. Hombach, R. Tay, and N. Kuster, "The Dependence of EM Energy Absorption upon Human Head Modeling at 1800 MHz," *IEEE Trans. Microwave Theory Tech.*, Vol. 45, pp. 2058–2062, October 1997.

38. O. Kivekas, J. Ollikainen, T. Lehtiniemi, and P. Vainikainen, "Bandwidth, SAR and Efficiency of Internal Mobile Phone Antennas," *IEEE Trans. Antennas Propagat.*, Vol. 46, pp. 71–86, February 2004.

39. J. C. Lin, "Specific Absorption Rates (SARs) Induced in Head Tissues by Microwave Radiation from Cell Phones," *IEEE Antennas Propagat. Mag.*, Vol. 42, pp. 138–139, October 2000.

40. A. Taflove, *Computational Electrodynamics: The Finite-Difference Time-Domain Method*. Boston: Artech House, 1995.

41. P. Silvester and R. Ferrari, *Finite Elements for Electrical Engineers, 2nd ed.* Cambridge, U.K.: Cambidge University Press, 1990.

42. W. Gander, *Computermathematik*. Birkhäuser, Basel, Zweite Auflage 1992.

43. R. G. Vaughan and N. L. Scott, "Evaluation of Antenna Configurations for Reduced Power Absorption in the Head," *IEEE Trans. Antennas Propagat.*, Vol. 48, pp. 1371–1380, September 1999.

44. J. Toftgard, S. N. Hornsleth, and J. B. Anderson, "Effects on Portable Antennas by the Presence of a Person," *IEEE Trans. Antennas Propagat.*, Vol. 41, pp. 739–746, June 1993.

45. L. C. Fung, S. W. Lenug, and K. H. Chan, "An Investigation of the SAR Reduction Methods in Mobile Phone Applications," *IEEE International Symposium on Antennas & Propagation*, San Antonio, TX. July 2002.

46. S. V. Amos, M. S. Smith, and D. Kitchener, "Modeling of Handset Interactions with the User and SAR Reduction Techniques," *IEE National Conf. Antenna Propagat.*, pp. 12–15, 1999.

47. M. Jung and B. Lee, "SAR Reduction for Mobile Phones Based on Analysis of EM Absorbing Material Characteristics," *IEEE International Symposium on Antennas & Propagation*, Columbus, OH. June 2003.

48. S. Eggleston, "Antenna Assembly, and Associated Method, Having Parasitic Element for Altering Antenna Pattern Characteristics," U.S. Patent No. 6249225, June 19, 2001.

49. M. Sager, M. Forcucci, and T. Kristensen, "A Novel Technique to Increase the Realized Efficiency of a Mobile Phone Antenna Placed Besides a Head-Phantom," *IEEE International Symposium on Antennas & Propagation*, Columbus, OH. June 2003.

50. R. Y. Tay, Q. Balzano, and N. Kuster, "Dipole Configurations with Strongly Improved Radiation Efficiency for Hand-Held Transceivers," *IEEE Trans. Antennas Propagat.*, Vol. 46, pp. 798–806, June 1998.

4

Wireless Channel Model

Massimo Franceschetti and Daniele Riccio

4.1 Introduction

In this chapter, we present an overview of problems encountered and procedures applied to estimate the electromagnetic field excited by antennas located in a built-up environment. We present model-based theories and procedures along with numerical, analytical, and experimental results. Model-based theories have the advantage, with respect to an exact approach, to provide feasible solutions and, with respect to any empirical approach, to provide more reliable solutions. Our presentation is divided into two logically different parts. The first part, presented in Section 4.2, provides the rationale for the design of numerical and graphical solvers for urban propagation, with one example of realization. The second part, presented in Section 4.3, is devoted to examples of analytical approaches based on probability theory that lead to average value solutions.

Propagation in the urban environment is, in principle, a well-posed problem: the electromagnetic field radiated by prescribed sources, in the simplest case a single antenna, must fulfill Maxwell equations in the open space and appropriate boundary conditions over all the built-up areas. However, any attempt to provide a solution to this apparently simple problem must face the complication of the scattering scenario. Each city is different from any other city, so general assessments are difficult to obtain. The type and distribution of buildings may drastically change among different sections of the same

161

city, so computational procedures must be changed accordingly. Even within the same section of a chosen city, the scattering scenario, i.e., the electromagnetic boundary conditions, may change in time, because the city is a living organism, with its cycles and temporal variations. Accordingly, each city may be considered an element of an ensemble, and the study of electromagnetic propagation and scattering may be pursued along essentially two lines of thought: either the deterministic or the stochastic approach. The deterministic approach makes reference to a single element of the ensemble, while the stochastic exploits the statistical properties of the distribution.

4.1.1   The Deterministic Approach

In the deterministic approach, an element of the cities ensemble is chosen, namely, the particular city of interest. This implies that knowledge of the three-dimensional (3D) geometry of the city must be known, i.e., shape and location of each building. Each building is then schematized in terms of a parallelepipedic structure topped with a (usually) flat roof; and average electromagnetic properties of the buildings' walls are either known or postulated. In the following, this model is referred to as the city Deterministic Geometrical Model (DGM). Ray tracing procedures are implemented to compute the electromagnetic field everywhere in the open space surrounding the buildings, i.e., in the streets and squares of the city. Sometimes, a rough estimate of the field inside the buildings is also obtained.

As stated above, field computation relies on ray tracing procedures: reflected rays are evaluated by using Fresnel reflection coefficients over the buildings' walls; and diffracted rays from the buildings' edges are accounted for either by the Geometrical Theory of Diffraction (GTD), with appropriate transition functions at the lit-shadow boundary, or by the Uniform Theory of Diffraction (UTD). More refined techniques also consider propagation over as well as through the buildings: the former with the aid of creeping rays sliding over the buildings' (flat) roofs, and the latter by applying semi-heuristic attenuation coefficients accounting for the building insertion within the ray path. The final result is the design of an electromagnetic solver, which provides numerical and graphical representation of the field intensity within the explored part of the city.

The presented approach exhibits a number of attractive features. Successive ray reflections may properly account for guided propagation along the city streets (canyon effect); and diffracted and creeping rays, as well as propagation through the buildings, may provide illumination in dark regions that are not reached by direct and successively reflected rays. As already mentioned, some rough estimate of the indoor field may be gained. Transient propagation, i.e., modulated signals, may be, in principle, accounted for by Fourier inversion of superposed frequency domain results.

However, this approach also suffers a number of limitations, model validity, adopted algorithm precision, and computational efficiency, which are examined in the following.

As already stated, the city DGM relies on the schematization of building as parallepipedic homogeneous dielectric structures, with plane walls and sharp edges, vertically standing over a homogeneous (not necessarily plane) background. It is clear that this is a rather rough idealized model of the real situation. The buildings' walls are usually corrugated with different types of openings (doors, windows) and a number of protruding features (balconies, decorations, etc.). The wall material may change in the same building. The street surface is neither homogeneous nor flat if pavements for walking are provided. In addition, parked cars, moving cars, and walking people are usually present in the streets, often together with trees, post, lamps, etc. The squares may be embellished with fountains and other types of decorations. All those objects are either moderately larger or of comparable size with respect to the operating wavelength, thus providing significant scattered field contributions that are ignored by the adopted model. These contributions are relevant, especially when the transmitting antennas are located at the street level (as in the case when cells dimensions are reduced), and are certainly significant even when the base stations are situated on the top of the buildings.

The city DGM tries to account for these deficiencies with heuristic procedures. The ray phase is supposed to undergo a random change when the ray hits each one of these additional obstacles. On one side, this is very crude because it neglects the real scattering properties of the obstacles, as their radiation diagrams and scattering cross section, and the associate ray spreading. On the other side, the attractive feature of the evaluation of fields interference is lost, due to their assumed random phases, and power averages must be considered. The alternative procedure of adding to the ray field an incoherent clutter component makes use of empirical expressions and suffers the same previous criticism.

Coming to the algorithm which is used, i.e., superposition of different species (direct, reflected, diffracted, creeping) of rays, it must be observed that rigorous diffraction coefficients are known only for perfectly conducting wedges, which is not the case in the city buildings. Accordingly, semi-heuristic expressions must be used, and the evaluation of diffracted and creeping rays turns out to be approximate.

As a last comment, the running times of these electromagnetic solvers are usually significant unless computer clusters are used, thus increasing implementation cost.

4.1.2 The Stochastic Approach

The philosophy which forms the basis of the probabilistic approach is completely different from that of the deterministic one. The deterministic approach considers just an element of the city ensemble and numerically computes the electromagnetic field there. The aim of the stochastic approach is to derive general analytical expressions, describing the *average* properties of the urban propagation. In addition, analytical results are required to contain a minimum number of physically meaningful parameters. Two probabilistic models are considered in the following.

The first model, the city Stochastic Environment Large-scatterers Model (SELM), is based on a probabilistic description of the buildings' distribution. Consider a two-dimensional (2D) regular array of rectangular (or squared) sites: each site may be either empty, with probability p, or occupied, with probability $q = 1 - p$. The state of each site is independent from that of all other sites. This arrangement is named *percolation lattice*, and it is a possible model for the cities stochastic distribution. The attractive feature of this model is the imprinting of the city's structure in the city's probabilistic description. There is, however, a significant limitation associated to the city SELM: the geometry is 2D, perhaps being more appropriate to low level transmitting antennas. Also, ray propagation ignores diffraction (which is instead included in numerical solvers): the canyon effect is obtained by successive reflections that allow rays' penetration along the two orthogonal directions of the lattice.

The second model, the city Stochastic Environment Small-scatterers Model (SESM), is the dual of the previous one: it is assumed that the scattering process is dominated by the myriad of small scatterers present in the city body and not by the large buildings. Although this assumption seems hard to accept at first glance, a little thought shows that it relies on a reasonable basis. There is no doubt that propagating rays are reflected (and diffracted) by the large building, but, in any case, not specularly because their walls are not smooth, as discussed in Section 4.2. In addition, the very large number of collisions of the rays with the myriads of small scatterers present in the city body, see Section 4.2, results in large spreading of their propagation direction, specific intensity, and polarization. In other words, the information that the ray underwent successive reflections is somehow lost, as the field propagates in the urban environment. In any case, the city SESM, considering only small obstacles, suffers limitations, and it is certainly interesting to make a comparison between SESM and SELM's predictions.

4.2   The Deterministic Geometrical Model (DGM)

This model is included in specific software packages that support wireless network design, identification of the (optimal) base station sites, and determination of antenna (optimal) input power, position, and orientation. More specifically, these tools can be used to find a solution to some basic questions relevant to the engineering side of the problem. A partial list includes increasing the signal-to-noise ratio; increasing the communication channel capacity; reducing the cost for unit bandwidth; and reducing, monitoring, and controlling electromagnetic compatibility, particularly with the population. In summary, these tools should lead to the maximization of service quality and efficiency, jointly with the minimization of required resources and cost.

   A deterministic solution can be, in principle, obtained by numerically solving Maxwell equations in the considered environment. However, due to the complexity of the environment, the computational time required is unacceptable. For this reason, it is convenient to cast the solution of the problem as the superposition of canonical elementary solutions regulating the interaction between the electromagnetic field and the scene. The choice of these canonical solutions is the key point to solve the problem in a feasible way. Recently, many computer tools based on ray tracing have been developed following this approach. Presenting a list of these tools is beyond the scope of this chapter. This list continuously changes, and features of each tool are updated. In the following, the rationale to design one of these tools is considered. This includes all the key features that are common to most of them. The tool design is divided into key items, and then results from the software tool EXACT, written in IDL (Interactive Data Language), are finally presented.[*]

The electromagnetic solver is applied to a prescribed situation referring to an actual antenna illuminating an actual urban area. In a generic urban area the terrain below the buildings is not flat and the buildings are not uniformly distributed. Moreover, it is evident that the buildings' orientation, position, form, and height largely influence the electromagnetic field propagation. Hence, a reliable tool should be based on an accurate description of the scene, usually implemented by means of a raster file that encodes the 3D map of the scene. The electromagnetic field can then be conveniently described as an appropriate ray congruence.[1] Within this framework, direct and inverse ray tracing procedures provide the appropriate asymptotic solutions to Maxwell equations that allow evaluation of any significant contribution to the electromagnetic field at any position of the scene.

The design of the software tool can be arranged in a four-step procedure, with each step consisting of finding convenient solutions to several smaller problems. The four steps, which are detailed in the subsequent sections, refer to (i) input data, (ii) output data, (iii) ray propagation, reflection, and diffraction, and (iv) outdoor/indoor contribution. Before discussing these steps, it is noted that ray tracing techniques also allow the time delay and spread of a radiated pulse to be evaluated by means of simple information on its optical path. It is therefore forecast that the future generation of these tools will include impulse response analysis with application to the network design.

4.2.1   Input Data

Input data refer to the electromagnetic field sources and the scene which sets the boundary conditions. The case of the transmitting antenna of a cellular base station in an urban environment is considered here. It is important to appropriately define the source and the scene input data because they implicitly define the class of problems that the tool can solve. Moreover, a clever arrangement of the input data is the key point to finding a reliable solution in acceptable computer running time.

---

[*] The tool used to generate the examples was designed at the University Federico II of Naples, Italy, and implemented by *WISE*, a small high-tech company located in Naples.

The source input data describe an actual antenna whose characteristics are appropriately defined. The antenna parameters that need to be included can be arranged into two groups: the *geometrical parameters* and the *electrical parameters*. The first group includes:

1. Geo-referenced 3D position
2. Mechanical tilt angle
3. Orientation toward the north direction

The second group includes:

1. Carrier frequency
2. Gain
3. Input power
4. Input resistance
5. Radiation patterns (at least onto two orthogonal planes)
6. Polarization

In Figure 4.1, a graphical representation visually summarizes these source input data for a given example.

Turning to the scene input data, these describe the site where propagation takes place. The terrain is represented by means of a raster file including terrain height at a prescribed spatial spacing (of the order of a few meters) and with a prescribed accuracy (of the order of 1 m). Random parameters are used to describe the terrain roughness at spatial scales smaller than the spatial spacing employed for the raster profile, but comparable to the employed electromagnetic wavelength. For each point of the terrain, the electromagnetic properties are described via the permittivity and conductivity that constitute the complex dielectric constant. The superimposed buildings are described as prisms and are geometrically characterized by the coordinates of the vertices of their top side. Each wall of the building is also described by the percentage of windowing. Moreover, knowledge of the thickness and complex dielectric constant of wall and glass is useful. Finally, a valuable planning tool should also accept, as input, any other ancillary or measured data that can be used to check the validity of the prediction.

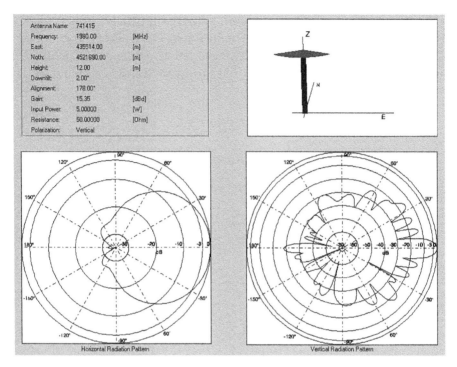

Figure 4.1 Source input data. Graphical representation of the antenna description appearing in a typical software solver.

The graphical input scene resulting from a raster file is depicted in Figure 4.2.

4.2.2 Output Data

In order to optimally design the communication network, the electromagnetic solver provides the estimated electromagnetic field on a regular spaced grid at a prescribed height over the terrain or the buildings' roof. The indoor field due to the outdoor antenna can also be evaluated for each building floor according to the same space sampling employed in the outdoor case.

It is convenient to operate in the phasor domain, thus evaluating each field component (in amplitude and phase) at given (x, y, z) coordinates or prescribed directions. This is important if polarization-dependent applications are considered.

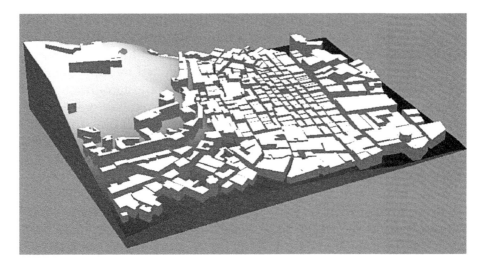

Figure 4.2 Scene input data. In this input scene representation, the antenna is depicted as a triangular spot on the top of the building placed on the top of the hill.

A way to somehow include the overall random nature of the urban environment can be implemented by adding to the computed field some random contributions. These contributions arise, for instance, because of vegetation and human-dependent elements like cars and people. Finally, some processing of computed field data is a valuable addition to the solver. For instance, the solver may also store results relevant to different antenna parameters so as to allow the estimation of the optimal antenna configuration.

4.2.3 Ray Propagation, Reflection, and Diffraction

In order to obtain the output data, geometrical optics are used to compute the electromagnetic field via a direct or inverse ray tracing approach.

Direct ray tracing, often termed ray launching, consists of evaluating the ray amplitude starting from the source antenna toward any possible direction up to any position in the scene. This allows the field intensity along the propagation path to be controlled. The procedure is stopped as soon as the ray intensity decreases under a prefixed sensitivity threshold. The ray propagation is terminated as soon as its intensity decreases under a prefixed threshold. The total field at any point is obtained by summing up the incoming rays therein. Conversely, in the inverse ray tracing approaches, the receiving position is first set. Source and receiver locations may be linked by several, in principle infinite, ray paths. These are obtained by implementing rays' reflections and diffractions from buildings and terrain. Then, all ray paths that link the source antenna and the receiver position are determined. This procedure is stopped by considering only those rays whose number of interactions with the buildings does not exceed a prescribed level. For global radio-coverage planning, the inverse ray tracing procedure is iterated for any considered location within the area. A simple criterion can be used to choose the best algorithm between direct and inverse ray tracing approaches. Inverse ray tracing can be conveniently employed whenever the electromagnetic field level must be evaluated only at some specific locations in the scene: in this case the procedure is efficient and feasible. Conversely, if global radio-coverage maps are of interest, the direct ray tracing approaches are preferable with respect to the inverse ones. Then, for the applications considered in this chapter, the software tool should implement direct ray tracing techniques. However, in this case, the computational time is extremely high, and appropriate algorithms must be introduced for a handy use of the tool, as discussed in the following. A dramatic reduction of the computational time for the direct ray tracing techniques is obtained if the ray launching procedure is not fully developed in the 3D environment. Assuming that all the buildings within the scene have vertical walls (i.e., orthogonal to the terrain), then the VPL (Vertical Plane Launching)[2] method can be employed without any approximation: the 3D ray launching is accomplished via a two-step 2D ray launching procedure: the first step is implemented in a single horizontal plane (HP); and the second step is implemented in a set of vertical planes (VPs). The HP coincides with the (x, y) plane; the VPs contain the z-axes and the source antenna, see Figure 4.3.

In the first step the HP is considered, see Figure 4.3(a). All rays launched from the source antenna and belonging to the same VP hold the same projection onto the HP. The buildings projections onto the HP consist of closed polygons. The projections of both the VP and the buildings onto the HP are analyzed, and the intersections between these two classes of projections are classified and stored on a computer. These intersections take place on the buildings' walls and may give rise to reflected rays. In addition, if the intersection is close enough to the wall boundary, then diffracted rays may originate. Intersections

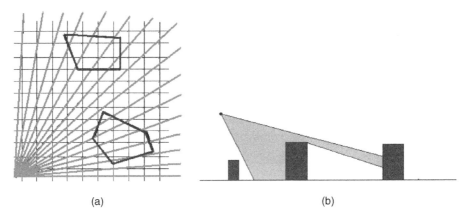

(a)   (b)

Figure 4.3 Vertical and horizontal planes. (a) Propagation in the HP. Projections of the VPs are lines. The antenna is positioned in the bottom-left corner. (b) Propagation in the VP. Buildings may fully (or partly) intercept the rays from the source antenna.

are possible sources of reflected and diffracted rays and are classified accordingly.

In the second step the VP is considered, see Figure 4.3(b). The existence of all the possible (diffracted or reflected) sources is verified onto the VP. As a matter of fact, all the rays belonging to the same VP travel along different ray paths and may hit the ground, the building wall, the roof, or even pass over all the buildings. Simple algorithms can be developed to verify all these events, and the corresponding results are classified accordingly. The effect of interaction with the buildings' walls and terrain, causing ray reflections, is evaluated according to the Geometrical Optics (GO) theory. Conversely, the effect of interaction with the walls' edges, causing ray diffraction, is evaluated according to the UTD. Diffractions from vertical buildings' edges (walls) as well as horizontal ones (roofs) are taken into account. As a side comment, note that the tool presented here takes into account all the rays diffracted by the horizontal edges, at variance with other VPL codes. Efficient algorithms have been developed to reduce the computational time. If a reflecting or a diffracting element is encountered, a virtual (image) source is created and stored. When all the rays from the primary source have been analyzed, a full set of new (image) sources is generated and stored on a computer. Then, the algorithm is iterated by analyzing each virtual source. Ray propagation is discarded when the associated field intensity reaches a prescribed minimum level. Hence, no limits are preliminarily imposed on the number of possible reflections and diffractions. The field evaluated at each point of the output grid is composed of the contributions of direct, reflected, and diffracted rays. Due to the random nature of the urban environment and the output pixel dimension, usually much greater than the electromagnetic wavelength, the field phases are uniformly distributed variables: an incoherent contributions sum is advisable to get the expected value of the power density. However, field summations can also be implemented for interference and fading estimates.

4.2.4 Results

We now present some results obtained using the software tool EXACT. The electric field map, evaluated for the input scenario presented in Figure 4.2, is reported in Figures 4.4 and 4.5.

The tool also computes the indoor field due to an outdoor antenna. The evaluation is performed by modelling each wall of the building as a homogeneous planar interface whose electromagnetic and thickness parameters are obtained by an averaging procedure accounting for the presence of different materials: bricks of the solid wall and glass of the windows. The average takes into account the percentage of windowing for each wall. Examples of the results are depicted in Figure 4.6, where the electric field map evaluated for each floor in a four-floor building included in the scene of Figure 4.2 is reported.

We also report some comparisons between predicted and measured data. Measured data were collected by means of a mobile receiving station mounted over a car and proceeding along four paths within the scene. Measured data were compared with the corresponding predicted ones at the same receiver locations. The absolute values of the differences between predicted and measured fields (in decibels) are depicted in Figure 4.7. These differences are of the order of magnitude which is expected for these kinds of software tools. They may be improved by better knowledge of electromagnetic parameters and by implementing appropriate averaging techniques.

4.3 The Stochastic Environment Model

The stochastic approach provides analytical solutions by exploiting two main simplifications. On one side, a stochastic model of a canonical city is provided, relying on only a few parameters. On the other

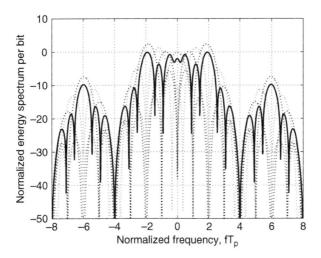

Plate 1 The average energy spectrum per bit for an OCDM spreading signal set defined in Figure 2.5. (See Figure 2.6, p.34)

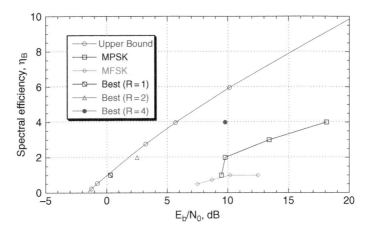

Plate 2 A comparison between the modulations considered in this chapter. (See Figure 2.10, p.39)

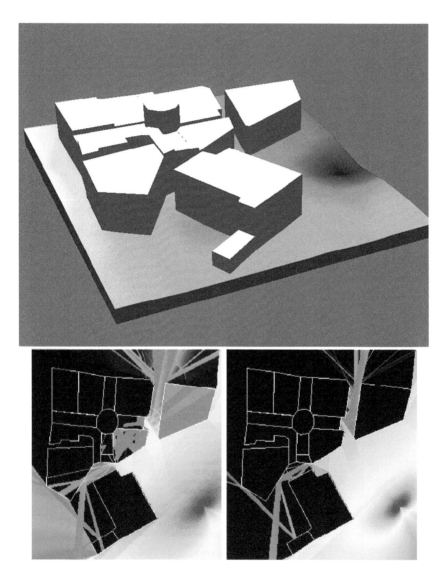

Plate 3 Field intensity evaluation. (Top) Simulation of the electric field in the portion of an urban area. (Bottom-left) Simulation of the electric field amplitude relevant to direct and reflected contributions. (Bottom-right) Simulation of the electric field amplitude relevant to direct, reflected, and diffracted contributions. The electric field amplitude is color coded between −25 and −75 dBm. (See Figure 4.4, p.173)

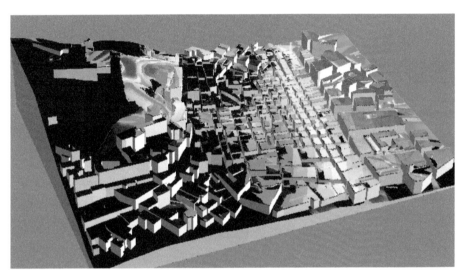

Plate 4 Field intensity evaluation. Simulation of the electric field for the scene in Figure 4.2. (See Figure 4.5, p.174)

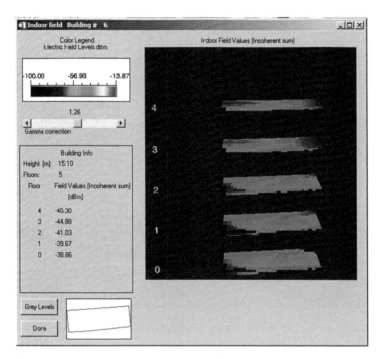

Plate 5 Indoor field intensity evaluation. Simulation of the electric field for each floor of a four-floor building included in the scene of Figure 4.2. The field intensity is color coded between −13.87 and −100 dBm. (See Figure 4.6, p.174)

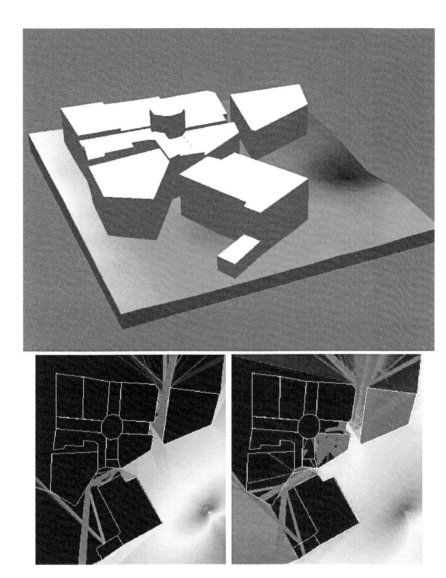

Figure 4.4 Field intensity evaluation. (Top) Simulation of the electric field in the portion of an urban area. (Bottom-left) Simulation of the electric field amplitude relevant to direct and reflected contributions. (Bottom-right) Simulation of the electric field amplitude relevant to direct, reflected, and diffracted contributions. The electric field amplitude is color coded between –25 and –75 dBm. (See Color Plate 3)

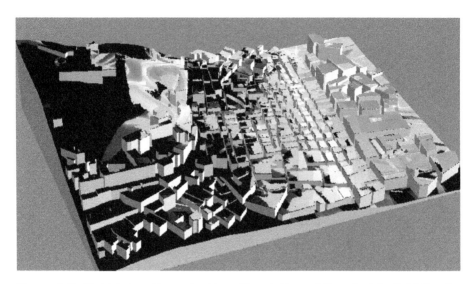

Figure 4.5 Field intensity evaluation. Simulation of the electric field for the scene in Figure 4.2. (See Color Plate 4)

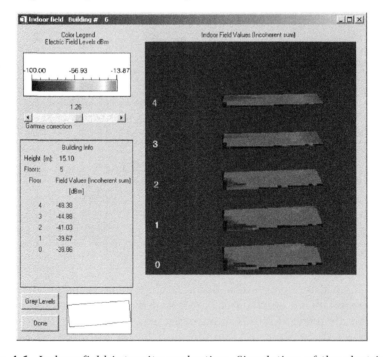

Figure 4.6 Indoor field intensity evaluation. Simulation of the electric field for each floor of a four-floor building included in the scene of Figure 4.2. The field intensity is color coded between –13.87 and –100 dBm. (See Color Plate 5)

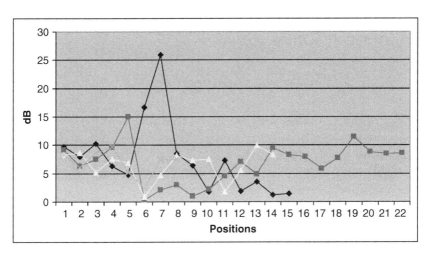

Figure 4.7 Prediction error. Plot of the difference between predicted and measured electric field amplitudes. Four data sets acquired along different paths are depicted.

side, simplifying assumptions are made on the propagation mechanism. We focus on two stochastic models that appeared in the literature.[3–7] In 3–7 these models were also compared with classic ones that have been extensively used to model propagation in the atmosphere.[8] Both models provide a simplified vision of the propagation mechanism, but also generate very relevant field predictions. We compare their different assumptions and discuss their range of applicability.

The first model (SELM) assumes that the typical dimension of the obstacles encountered by the propagating wave is large compared to the wavelength of the radiated waveform. This implies that rays that propagate in a random environment following the simple Snell law of reflection (specular obstacles). The second model (SESM) assumes that the obstacles are small compared to the transmitted wavelength, this implies that individual photons are scattered by the obstacles, each in a random direction (diffusive obstacles). Accordingly, in the first case, electromagnetic (EM) energy is conveyed along canyons formed between obstacles that continuously leak (and collect) energy along their lateral openings. In the second case, EM energy is diffused in the environment where each point scatterer is effectively a point source of a new wave. In other words, while the first model assumes that the fundamental contribution to EM propagation in the environment is due to multiple reflections, the second model assumes that the fundamental contribution is due to multiple scattering. Hence, the first model is more suited to describe propagation in urban areas populated by high rise buildings, while the second model is more suited to environments rich in vegetation or urban paraphernalia.

4.3.1 The Large-Scatterers Model (SELM)

The first model we consider first appeared in Reference 3, where the idea of representing the urban propagation channel as a random lattice was exploited for the first time.

Consider a regular lattice of square sites that can be either occupied with probability q or empty with probability $p = 1 - q$. A key observation is that a picture of an urban area taken from an aircraft would not be so different from the one depicted in Figure 4.8, where occupied sites of the lattice represent buildings, the model parameter q is the building density of the city, and EM propagation occurs along canyons formed between clusters of occupied sites. It is assumed that rays propagate inside the lattice following the simple laws of specular reflection, where at each reflection a portion of the EM energy is absorbed by the obstacles. Hence, the two parameters characterizing this model are the obstacles density q and their reflection coefficient $\xi < 1$.

Figure 4.8 SELM. A percolative lattice with some rays depicted. The rays depart isotropically from the EM source located at the center of the shown lattice.

What we have described is certainly an oversimplified model of EM propagation in urban areas for many reasons. These include neglecting diffraction effects on building corners and roofs, addressing only 2D propagation, ignoring trapping/absorption effects of the lattice (e.g., open doors or windows), etc. In spite of this, it can be used to determine to what extent the presence of a regular (although stochastic) background in built-up areas imprints its presence on the received signal and can lead to very relevant field predictions.

4.3.2   The Small-Scatterers Model (SESM)

The second model we consider first appeared in Reference 4, where the idea of modelling propagation using continuous random walks was exploited for the first time.

We refer to Figure 4.9. The propagating environment is modelled as a random distribution of point scatterers, and the propagating wave is modelled as an ensemble of photons that spread in the space, may hit the scattering objects, and are scattered around. Accordingly, each photon propagates along a piecewise linear trajectory, where the turning angle at each hit is uniformly chosen at random, and the average random length of the line segments characterizes the obstacles density.

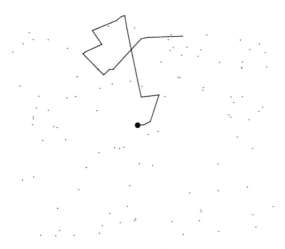

Figure 4.9 SESM. Each photon radiating from an isotropic source propagates in the environment following a piecewise linear trajectory, modelled as a random walk. Each time the photon hits an obstacle, it turns in a random direction or is absorbed by the obstacle.

At each hit, the propagating photon may be absorbed, with some probability, by the obstacle, or it may be scattered in a random direction. Hence, the two characterizing parameters are in this case the average step length of the random walk, representing the amount of clutter in the environment, and the absorption probability $\gamma < 1$.

Also, in this case we oversimplified the real propagation mechanism. However, we point out that when low level antennas are employed, like in microcells for personal communication and in networks of multi-hop wireless sensors and laptop computers, multiple scattering by diffusive objects plays a key role. In outdoor areas, lamp posts, street signs, trees and vegetation, pedestrians, cars, and irregularly sited and textured building walls can be in first approximation, considered diffusive, as point scatterers. Similarly, indoor scattering is provided by the presence of furniture, people, textured walls, doors, etc.

4.3.3 Analytic Results

In this section we summarize analytic results that were obtained for the two models. The first set of results was obtained in Reference 3, where simple formulas for the penetration capability of a plane wave (an ensemble of parallel rays) impinging at a given angle θ on a half-plane lattice were given. This is a canonical scenario representing a transmitting antenna placed at the boundary of a city, see Figure 4.10. Obtained formulas, valid for uniform building distribution of constant parameter $p = 1 - q$, can be generalized to the case when the parameter $p_j = 1 - q_j$ is a function of the penetration index j that indicates the number of lattice levels along the positive x-axis. For example, when modelling a city as a discrete grid, the occupation probability would increase as we approach the city center. The result is the following. The probability that a ray incident at angle θ reaches level k inside the half-plane lattice before being expelled out of the lattice is given by

$$P(r_N > k) \approx \frac{p_1}{k} \sum_{i=1}^{k-1} iq_{ei} \prod_{j=1}^{i-1} p_{ej} + p_1 \prod_{j=1}^{k-1} p_{ej}, \qquad (4.1)$$

where

$$p_{ej} = p_j^{\tan\theta} p_{j+1},$$
$$N = \min\{n : r_n \geq k \text{ or } r_n \leq 0\},$$

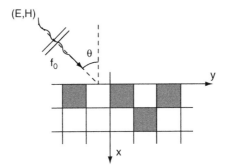

Figure 4.10   Canonical scenario, SELM. A plane wave is incident at a certain angle θ on a half-plane random lattice.

and r_n is the ray's position after n reflections. Note that the obtained formula (4.1) depends only on the states of the cells up to level $k-1$ and is independent of the state of the other cells, as expected on physical grounds. Moreover, in the case $p_j = p\ \forall j$, this reduces to the formula given in[3]

$$P(r_N > k) \approx \frac{p}{q_e^k}(1 - p_e^k),\qquad(4.2)$$

where

$$p_e = 1 - q_e = p^{\tan\theta+1}.$$

The above formulas were obtained under the simplifying assumption of independence of successive hits. In practice, this corresponds to assuming the incident ray to explore new portions of the lattice as it propagates inside it, and numerical simulations show that this is a valid approximation when the incident angle is close to $45°$.[3] Work is in progress to relax this limitation. The formulas can thus be used to evaluate the coverage by an antenna of a given urban area of building distribution profile p_j, $j = 1, 2 \ldots$, at any layer depth k.

However, when the source is placed inside the lattice, as depicted in Figure 4.8, all propagating angles, not only those close to $45°$, must be considered and above derivations do not hold. A somehow less formal solution is obtained in Reference 7 in the case of uniform building distribution and using a maximum entropy approach. Let us consider the probability $Q_n(m, k)$ of the ray undergoing its nth reflection at Manhattan distance $|m| + |k|$ from a source placed at coordinates $(0, 0)$ inside the lattice. We write this distribution as the one exhibiting the maximum Shannon's entropy among all distributions satisfying a constraint on the average Manhattan displacement after n steps $D_n = E_n(|m| + |k|)$. Assuming anomalous diffusion of the rays, $D_n = \alpha n^\beta$, where $\beta < 1/2$ is the anomalous diffusion exponent, and α is a given constant, we obtain

$$Q_n(m, k) \approx \frac{1}{\alpha^2 k^{2\beta}} \exp \left\{ -\frac{2(|m| + |k|)}{\alpha k^\beta} \right\}. \tag{4.3}$$

Equation (4.3) can then be used to compute the average path loss inside the medium as

$$\langle PL(m, k) \rangle = \sum_n \alpha^n Q_n(m, k) \approx \frac{e^{-b(|m| + |k|) \frac{1}{\beta+1}}}{(|m| + |k|)^d}, \tag{4.4}$$

with b and d given constants dependent on the reflectivity of the obstacles and on the dimensionality of the problem. We come back to this formula in the next section. First, we describe results obtained for the SESM.

In this second model each radiating photon proceeds along a straight line for a random length until it hits an obstacle. The photon is then either absorbed by the obstacle, with probability γ, or is scattered in a random direction, chosen uniformly in $[0, 2\pi]$, with probability $1 - \gamma$. In this way, a photon propagates in the environment along a random piecewise linear trajectory (see Hughes[9] for background on continuous random walks), as depicted in Figure 4.9. Let $Q(r)$ be the probability density function (*pdf*) of hitting an obstacle at distance r from the photon's source, at the first step of the random walk. The *pdf* $G(r)$ of the photon's eventual absorption, at distance r from the source, can be obtained by solving the following equation:

$$G(r) = \gamma Q(r) + (1 - \gamma) Q * G(r), \tag{4.5}$$

where the symbol "$*$" stands for convolution. The use of (4.5) can be justified as follows. Let us iteratively substitute the expression for $G(r)$ in the convolution integral. We have $G(r) = g_0(r) + g_1(r) + g_2(r) + \cdots$, with $g_0(r) = \gamma Q(r)$, $g_1(r) = (1 - \gamma) Q(r) * g_0(r)$, $g_2(r) = (1 - \gamma) Q(r) * (1 - \gamma) Q(r) * g_0(r), \ldots$. In this recursive series $g_0(r)$ describes the event that the photon is absorbed at the first step of the random walk, hitting a single obstacle; $g_1(r)$ describes the event that the photon is absorbed at the second collision; and so on. Since all these events are disjoint, we sum all the g_i's to obtain the overall *pdf* $G(r)$ of the photon's absorption at distance r from the origin.

Equation (4.5) can be easily solved in the Fourier domain. Let

$$g(\omega) = FT[G(r)], \ q(\omega) = FT[Q(r)]$$

be the Fourier transforms (FT) of $G(r)$ and $Q(r)$, respectively. We can then solve (4.5) algebraically in the Fourier domain, obtaining

$$g(\omega) = \frac{\gamma q(\omega)}{1 - (1 - \gamma)q(\omega)}, \tag{4.6}$$

so that

$$G(r) = FT^{-1}\left[\frac{\gamma q(\omega)}{1 - (1 - \gamma)q(\omega)}\right]. \tag{4.7}$$

Assuming the obstacles to be uniformly distributed according to a Poisson point process, the *pdf* $Q(r)$ is exponential of parameter η, that is an estimate of the density of the obstacles in the environment, and we can solve (4.7) and relate it to the path loss of the channel in terms of the Poynting vector intensity $S(r)$ and full power density $P(r)$, (see Franceschetti, Bruck, and Schulman[4]). In both cases expressions reduce (for SL propagation) to $1/(4\pi r^2)$ in the near field ($\eta r \ll 1$) and to a term proportional to $e^{-\beta \eta r}/r^2$ in the far field ($\eta r \gg 1$). This suggests using a simplified path loss formula of the type

$$\langle PL(r) \rangle = \frac{e^{-br}}{r^d}, \tag{4.8}$$

where b and d are constant values again dependent on the absorption coefficient and on the dimensionality of the problem ($d = 2$ in three dimensions).

The exponential form of (4.8) can be compared with that of (4.4). Both stochastic models predict an exponential power attenuation in the far range, due to absorption in the environment. The main difference is in the functional form of the exponential that in (4.4) depends on the Manhattan distance from the origin, while in (4.8) depends on the Euclidian distance. This is not surprising, given that the models describe two different scenarios: of large squared obstacles that reflect the incident wave and of small shapeless obstacles that uniformly diffuse the impinging wave. Nevertheless, both solutions lead to similar results of exponential or sub-exponential attenuation of the received power.

Finally, we point out that the case of a radiated pulse, rather than a monochromatic wave, is considered in Franceschetti[5] and leads to a similar exponential decay of the received power profile of the pulsed waveform in both time and distance from the transmitter.

4.3.4 Validation

We now compare predictions of the models with real data to assess the validity of the models. We use data[4] collected in the city of Rome, Italy, with a mobile receiving station (van). Details on the collection procedure are given in Franceschetti.[4] Figure 4.11 shows the raw data along the path of the receiving station, with the grey level proportional to the intensity of the received field. In order to assess the validity of the SELM, we have oriented the map along the (x, y) directions of the percolation model. The orientation cannot be perfect, due to the fact that in Rome the roads are not perfectly perpendicular. Curves at constant city block distances are squares centered at the origin and tilted by 45°. In order to obtain the averaged data point at a given Manhattan distance, we considered a family of such squares, for increasing values of $|x| + |y|$ and averaged the corresponding measured data along the squares, perimeter. After plotting the data as a function

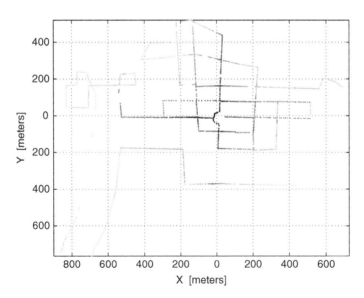

Figure 4.11 Collected raw data. As the van drove around the streets of Rome, it collected field values that are plotted with a grey level proportional to their intensity.

of the Manhattan distance from the origin, we performed curve fitting using (4.4). The result is depicted in Figure 4.12.

Similarly, after plotting the data as a function of the Euclidian distance from the origin (generating data samples along the radial direction by averaging on circles of increasing radii centered at the origin), we performed curve fitting using (4.8). The result is depicted in Figure 4.13 along with a fit performed using Hata's formula. This is an empirical formula, not supported by a theoretical model, which predicts an attenuation proportional to a power of the distance to the transmitter of the type $1/r^\alpha$ and has been quite popular among communication companies in the 1980s and 1990s, for quick prediction purposes.

The standard deviations of the models were $\sigma_{SELM} = 3.19$ dB and $\sigma_{SESM} = 3.02$ dB. Furthermore, a lower bound on the smallest achievable standard deviation obtainable with any monotonic function fitting the given data has been computed for the two cases using isotonic regression: $\sigma_{IsoSELM} = 1.43$ dB and $\sigma_{IsoSESM} = 1.37$ dB. The slightly better performance of the random walk model can be justified by the measurements being performed in a small cell, while the effect of imprint of the Manhattan geometry on the radiated field would be more evident at larger distances that, for the considered low-power experiment, resulted in a round-off to negligible values of the received

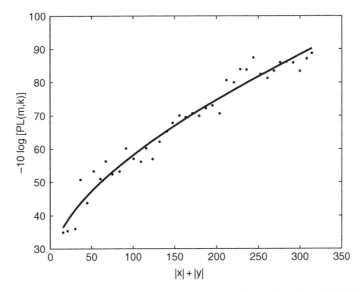

Figure 4.12   SELM fit. The path loss formula predicted by the SELM is plotted against the measured data.

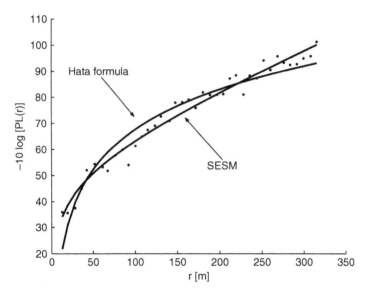

Figure 4.13 SESM fit. The path loss formula predicted by the random walk model is plotted against the measured data. Note also how this improves upon Hata's formula.[10]

field. In any case both models improve on the classical Hata's formula which has a standard deviation of 4.62 dB, overestimating the attenuation in the short range and underestimating it in the far field, as observed in Figure 4.13.

4.4   Conclusion

The problem of wave propagation in complex environments can be successfully attacked using a deterministic or stochastic modelling approach. We have summarized advantages and limitations of the two approaches, comparing predictions with real measured data.

The deterministic approach leads to the design and use of electromagnetic solvers that compute field intensity everywhere in any specified city. A geometric description of the urban environment is required, along with its electromagnetic properties. However, this information is either expensive or difficult to obtain, and this is the main problem for a practical use of the solvers.

The stochastic approach provides simple expressions to compute average decay of the field with the distance from the transmitter.

Obtained formulas are characterized by a few parameters representing average properties of a city. The choice of these parameters is the key problem for a practical usage of these expressions. Their successful definition and measurement can make this approach valuable.

This chapter provides an overview of the problem, indicating that an integrated approach of both deterministic and stochastic techniques may provide a deeper insight into the urban area wireless channel.

References

1. G. Franceschetti. *Electromagnetics. Theory, Techniques, and Engineering Paradigms*. Plenum Press, New York, 1997.
2. G. Liang and H. L. Bertoni. "A new approach to 3-D ray tracing for propagation in cities," *IEEE Trans. Antennas Propagat.*, 46, 853–863, June 1998.
3. G. Franceschetti, S. Marano, and F. Palmieri. "Propagation without wave equation towards an urban area model," *IEEE Trans. Antennas Propagat.*, 47(9), 1393–1404, September 1999.
4. M. Franceschetti, J. Bruck, and L. Schulman. "A random walk model of wave propagation," *IEEE Trans. Antennas Propagat.*, 52(5), 1304–1317, May 2004.
5. M. Franceschetti. "Stochastic rays pulse propagation," *IEEE Trans. Antennas Propagat.*, 52(10), 2742–2752, October 2004.
6. S. Marano, F. Palmieri, and G. Franceschetti. "Statistical characterization of ray propagation in a random lattice," *JOSA A*, 16(10), 2459–2464, 1999.
7. S. Marano and M. Franceschetti. "Ray propagation in a random latice: A maximum entropy, anomalous diffusion process," *IEEE Trans. Antennas Propagat.*, 53(6), 1888–1896, June 2005.
8. A. Ishimaru. *Wave Propagation and Scattering in Random Media. Vols. 1,2*. Academic Press, San Diego, 1978.
9. B. Hughes. *Random Walks and Random Environments. Vol. 1, Random Walks*. Oxford University Press, London, 1995.
10. M. Hata. "Empirical formula for propagation loss in land mobile radio services," *IEEE Trans. Vehicular Technol.*, 29(3), 317–325, 1980.

5

Ad Hoc Wireless Networks

Mario Gerla

5.1 Introduction and Definitions

An "ad hoc" network is a network established for a special, often extemporaneous, service customized to a particular application. In this network, any node is typically a participant for only a limited period of time. The protocols are tuned to the particular application (e.g., send a video stream across the battlefield, find out if a fire has started in the forest, or establish a videoconference among teams engaged in a rescue effort). The application may be mobile, and the environment may change dynamically. Consequently, the ad hoc protocols must self-configure to adjust to the environment, traffic, and mission changes. What emerges from these characteristics is the vision of an extremely flexible and malleable, yet robust and formidable network architecture. This architecture can be used to monitor birds in their natural habitat, to interconnect collaborating scientists during field explorations, or to launch deadly attacks onto unsuspecting enemies.

5.1.1 Wireless Evolution

We begin by studying the evolution of wireless communications systems and networks. The rapid advances of radio technology in the last 30 years have stimulated the development of mobile communications systems that meet the needs of professionals on the move. First, came the need to communicate while on the move or away from a fixed phone outlet or Internet plug. The First Wave, the cellular phone, took the original developers by surprise, but it was actually a very predictable phenomenon because telephony is by definition a mobile application, as are voice communications. In fact, most phone calls are made while we are away from home or the office, calling friends to coordinate our movements/meetings, asking for directions, etc.

Next, came the Second Wave in the form of the requirement to connect to the Internet from mobile terminals. The traditional Internet applications are less "mobile" than telephony (most of us would prefer to read our e-mail from the comfort of our home rather than from the road). However, since we are spending an increasing number of hours in cars, trains, and planes, we want to fully utilize the travel time by engaging in Internet work. Leveraging this trend, new emerging Internet "location-based" services (e.g., navigation assistance, store price comparisons, tourist/hotel/parking, etc.) will soon make the wireless Personal Data Assistant (PDA) or Smart Phone an indispensable companion. The Second Wave (mobile Internet access) is supported by wireless Local Area Network (LAN) technology (predominantly, based on the IEEE 802.11 standard) and by data cellular services (e.g., Generalized Packet Radio System [GPRS]; 1xRTT, Code Division Multiple Access [CDMA] based single carrier Radio Transmission Technology; and the emerging Universal Mobile Telecommunication System [UMTS], also known as third Generation Wireless [3G]). Both cellular and wireless networking services are supported by an infrastructure; they address well-established and understood "commodity" needs of the users (e.g., conferencing, e-mail, web access, etc.).

The Third Wave in this wireless revolution is represented by the "ad hoc networking" technology. This type of network was born with goals very different from mobile telephony and Internet access. The primary goal was to set up communications for specialized, customized, extemporaneous applications in areas where there was no preexisting infrastructure (e.g., jungle explorations, battlefield), where the infrastructure failed (e.g., earthquake rescue), or where it was not adequate for the current needs (e.g., interconnection of low energy environmental sensors). With the exception of environment sensor networks (where ad hoc networking is fixed and is motivated by a lack of convenient, low cost infrastructure), most of the other ad hoc applications are mobile. In fact, they often reflect coordinated mobility patterns (e.g., group motion, swarming, etc.). They involve heterogeneous node types (with different form, energy, transmission range, and bandwidth factors) and heterogeneous traffic (voice, data, and multimedia). They often pose critical time constraints (because of the multimedia traffic and the emergency nature of the applications).

In truth, the concept of ad hoc wireless networking was first introduced in the early 1970s under the name of "packet radio". It was then a technology that looked very appealing to the military to interconnect mobile assets in the battlefield. It was, in fact, the natural extension to wireless of the Packet Switching concept, successfully demonstrated in the late 1960s under another Army project, the Advanced Research Projects Agency Network (ARPANET). Military applications of the ad hoc networking technology have been around for over 30 years. Commercial applications, however, are still waiting to take off in the same explosive way as cellular telephony and Wi-Fi. So, when we refer to ad hoc networking as the "Third Wave," we should think more of a "surge" that has been building over the past several years and will show its full strength in the next few years.

5.1.2   Ad Hoc Network Characteristics

What are the unique properties of ad hoc nets that set them apart from other wireless networks? In this section we review the most important features of this new technology.

- **Mobility**: The fact that nodes move is the *raison d'etre* of ad hoc networks. Rapid deployment in areas with no infrastructure implies that the users must explore an area and perhaps form teams/swarms that, in turn, coordinate among themselves to create a task force or a mission. We can have individual random mobility, group mobility, motion along preplanned routes, etc. The mobility model can have a major impact on the selection of a routing scheme and can thus influence performance. Mobility has been traditionally the "enemy" of the network designer, in many ways making the design and operation very difficult. However, mobility can also help, for instance, in permitting opportunistic exchange of messages among roaming nodes as they *rendezvous* or in providing connectivity among isolated nodes using a mobile, aerial backbone network.

- **Multihopping**: A multihop network is a network where the path from source to destination traverses several other nodes. While multihopping is not a prerequisite for a network to be called ad hoc, ad hoc nets often exhibit multiple hops for obstacle negotiation, spectrum reuse, and energy conservation. Battlefield covert operations also favor a sequence of short hops with low transmit power rather than a long, high power hop to reduce detection by the enemy.

- **Self-organization**: The ad hoc network must autonomously determine its own configuration parameters, including addressing, routing, clustering, position identification, power control, etc. In some cases, special nodes (e.g., mobile backbone nodes) can self-organize and coordinate their motion throughout the network to provide coverage of disconnected islands

- **Energy conservation**: Most ad hoc nodes (e.g., laptops, PDAs, sensors, etc.) have a limited power supply and limited capability to generate their own power (e.g., solar panels). Energy efficient protocol design (e.g., Media Access Control [MAC], routing, resource discovery, etc.) is thus critical for longevity of the mission and is an important requirement in ad hoc network design.

- **Scalability**: In some applications (e.g., large environmental sensor fabrics, battlefield deployments, urban vehicle grids, etc.) the ad hoc network can grow to several thousand nodes. For wireless network infrastructures, networks scalability is simply handled by a hierarchical construction. The limited mobility of infrastructure networks can also be easily handled using Mobile Internet Protocol (IP) or local handoff techniques. Because of the more extensive mobility and the lack of fixed references, pure ad hoc networks do not tolerate a fixed hierarchy structure and do not scale well with extensive Mobile IP deployments. Thus, mobility is what makes large scale one of the most critical challenges in ad hoc design.

- **Security**: The challenges of wireless security are well known — the ability of the intruders to eavesdrop and jam/spoof the channel. A lot of the work done in general wireless infrastructure networks extends to the ad hoc domain. The ad hoc networks are even more vulnerable to attacks than the infrastructure counterparts. There are both active and passive attacks. An active attacker tends to disrupt operations (say, an impostor posing as a legitimate node intercepts control and data packets, reintroduces bogus control packets, damages the routing tables beyond repair, unleashes denial of service attacks, etc.). Due to the complexity of the ad hoc network protocols, these active attacks are by far more difficult to detect/fold in ad hoc than infrastructure nets. Passive attacks are unique to ad hoc nets. They can be even more insidious than the active counterpart.

The active attacker must expose himself and is eventually discovered and physically disabled/eliminated. The passive attacker is never discovered by the network. Like a "bug," it is placed in a sensor field or at a street corner. It monitors data and controls traffic patterns and thus infers motion of troops in the field or the evolution of a particular mission. This information is relayed back to the enemy headquarters via low energy, low probability of detect networks. Defense from passive attacks requires powerful novel encryption techniques coupled with careful network protocol designs.

- **Unmanned, autonomous vehicles**: Some of the popular ad hoc network applications require unmanned, robotic components. All nodes in a generic network are, of course, capable of autonomous networking. When autonomous mobility is also added, there arise some very interesting opportunities for combined networking and motion. For example, ground or airborne nodes (Unmanned Ground Vehicles [UGVs] and Unmanned Airborne Vehicles [UAVs]) cooperate in maintaining a large ground ad hoc network interconnected in spite of physical obstacles, propagation channel irregularities, and enemy jamming. If some applications are delay tolerant, the robots can "carry" data to a destination when a connected path does not exist (i.e., Delay Tolerant Networking).

- **Multihop connection to the Internet**: There is merit in extending the single hop wireless infrastructure opportunistically with multihop ad hoc appendices. For instance, the reach of a domestic wireless LAN can be extended as needed (to the garage, the car parked in the street, the neighbor's home, etc.) with portable routers. These opportunistic extensions are becoming increasingly important for commercial applications. The integration of ad hoc protocols with infrastructure standards is becoming a hot issue.

5.1.3 Wireless Network Taxonomy

From the above, it is clear that ad hoc nets offer challenges (and opportunities) well beyond the reach of infrastructure networks. So, where do these nets fit in the overall wireless network classification? Most researchers will view ad hoc networks as a specialized subset of wireless networks. In fact, the ad hoc radio technology will be driven by (and will benefit from) the advancements in "infrastructure" wireless LANs.

New radio technologies (e.g., Orthogonal Frequency Division Multiplexing (OFDM), Multi-Input Multi-Output (MIMO) antenna systems, and cognitive radios) will be introduced first in wireless LANs and then transferred to ad hoc nets. There are unique design features in ad hoc nets that mark a departure from wireless LANs, mostly in the network areas (addressing, routing, multicast, mobility support, network level security, etc.). To complete the wireless network landscape, sensor nets can be viewed, from the network standpoint, as a subset of ad hoc networks. At the physical, MAC, and network layers, the major innovations and unique features of sensor nets (which set them apart from conventional ad hoc networks) are the miniaturization, the embedding in the application contexts, and the compliance with extreme energy constraints. At the application layer, the most unique feature of sensor nets is undoubtedly the integration of transport and in-network processing of the sensed data.

5.2   Ad Hoc Network Applications

Identifying the emerging commercial applications of the ad hoc network technology has always been an elusive proposition at best. Of the three above-mentioned wireless technologies — cellular telephony, wireless Internet, and ad hoc networks — it is indeed the ad hoc network technology that has been the slowest to materialize, at least in the commercial domain. This is quite surprising since the concept of ad hoc wireless networking was born in the early 1970s, just months after the successful deployment of the ARPANET, when the military discovered the potential of wireless packet switching. Packet radio systems were deployed much earlier than any cellular and wireless LAN technology. The old folks may still remember that when Bob Metcalf (Xerox Park) invented the Ethernet in 1976, word spread that he rediscovered a way to "packet radio" on a cable!

Why so slow a progress in the development and deployment of commercial ad hoc applications? The main reason is that the original applications scenarios were NOT directed to mass users. In fact, to this day, the driving application is instant deployment in an unfriendly, remote, infrastructure-less area. The battlefield, Mars explorations, disaster recovery, etc. have been an ideal match for those features. Early Defense Advanced Research Project Agency (DARPA) packet radio scenarios were consistently featuring dismounted soldiers, tanks, and ambulances. A recent extension of the battlefield is the homeland security scenario, where unmanned vehicles (UGVs and UAVs) are rapidly deployed in urban areas hostile to man, say, to establish communications before sending in the agents and medical emergency personnel.

Recently, however, an important new concept has emerged which may help extend ad hoc networking to commercial applications, namely, the concept of an **opportunistic ad hoc wireless extension**. This new trend has been prompted by the popularity of wireless telephony and wireless LANs and, at the same time, by the recognition that the latter techniques have their limits. The ad hoc network is used "opportunistically" to extend a home or campus network to areas not easily reached by the above or to tie together Internet islands when the infrastructure is cut into pieces, for example, by natural forces or terrorists.

Another important area that has propelled the ad hoc concept is sensor nets. **Sensor nets** combine transport and processing and amplify the need for low energy operation, low form factor, and low cost. In a sense, these are specialized ad hoc networks. However, they have been growing much faster than ad hoc networks because they do have compelling applications.

In the following we elaborate on two key applications: the **battlefield** and the **opportunistic ad hoc extension**.

5.2.1   The Battlefield

In future battlefield operations, autonomous agents Unmanned Vehicles (UVs), some ground-based UGVs and some airborne UAVs will be projected to the forefront for intelligence, surveillance, strike, enemy antiaircraft suppression, damage assessment, search and rescue, and other tactical operations. The agents will be organized in clusters (teams) of small unmanned ground, sea, and airborne vehicles in order to launch complex missions that comprise several such teams. Examples of missions include coordinated aerial sweep of vast urban/suburban areas to track suspects, search and rescue operations in unfriendly areas (e.g., chemical spills, fires, etc.), exploration of remote planets, reconnaissance of an enemy field in the battle theater, etc. In those applications, many different types of Unmanned Vehicles (UVs) will be required, each equipped with different sensor, video reconnaissance, communications support, and weapon functions. A UV team may be homogeneous (e.g., all sensor UVs) or heterogeneous (i.e., weapon carrying UVs intermixed with reconnaissance UVs etc.). Moreover, some teams may be airborne, and other teams may be ground, sea, and possibly underwater based. As the mission evolves, teams are reconfigured, and individual UVs move from one team to another to meet dynamically changing requirements. In fact, missions will be empowered with an increasing degree of autonomy. For instance, multiple UV teams collectively will determine the best way to sweep a mine field or the best strategy to eliminate an air defense system. The successful, distributed management of the mission will require efficient, reliable, low latency communications within members of each team, across teams, and to a manned command post. In particular, future naval missions at sea or shore will require effective and intelligent utilization of real-time information and sensory data to assess unpredictable situations; identify and track hostile targets; make rapid decisions; and robustly influence, control, and monitor various aspects of the theater of operation. Littoral missions are expected to be highly dynamic and unpredictable. Communication interruption and delay are likely, and active deception and jamming are anticipated.

The U.S. Armed Forces are currently investigating efficient system solutions to address the above problems. UVs have proven to be valuable in gathering tactical intelligence by surveillance of the battlefield. For example, UAVs such as Predator and Global Hawk are rapidly becoming an integral part of military surveillance and reconnaissance operations. The goal is to expand the UAV operational capabilities to include not only surveillance and reconnaissance, but also strike and support missions (e.g., command, control, and communications in the battle space). This new class of autonomous vehicles is foreseen as being intelligent, collaborative, recoverable, and highly maneuverable in support of future naval operations.[5,47]

In a complex and large-scale system of unmanned agents, such as that designed to handle a battlefield scenario, a terrorist attack situation, or a nuclear disaster, there may be several missions going on simultaneously in the same theater. A particular mission is "embedded" in a much larger "system of systems." In such a large-scale scenario the wireless, ad hoc communications among the teams are supported by a global network infrastructure, the "Internet in the Sky" (see Figure 5.1). The global network is provisioned independently of the missions themselves, but it can opportunistically use several of the missions' assets (ground, sea, or airborne) to maintain ultimate connectivity.

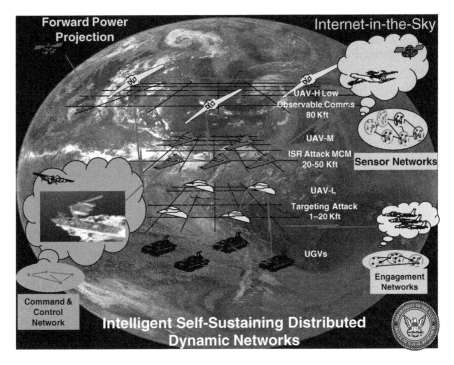

Figure 5.1   Autonomous agents with varying domains of responsibility.

The development of the Internet in the Sky for a military, battlefield scenario hinges on three essential technologies:

(a) Robust wireless connectivity and dynamic networking of autonomous UVs and agents

(b) Intelligent agents including mobile codes, distributed databases and libraries, robots, intelligent routers, control protocols, dynamic services, semantic brokers, and message-passing entities

(c) Decentralized hierarchical agent-based organization

As Figure 5.1 illustrates, the autonomous agents have varying domains of responsibility at different levels of the hierarchy. For example, clusters of UAVs operating at low altitude (1–20K feet) may perform combat missions with a focus on target identification, combat support, and close-in weapons deployment. Mid-altitude clusters (20–50K feet) execute knowledge acquisition, for example, surveillance and reconnaissance missions such as detecting objects of interest, performing sensor fusion/integration, coordinating low altitude vehicle deployments, supporting and medium-range weapons. The high altitude cluster(s) (50–80K feet) provides the connectivity. At this layer, the cluster(s) has a wide view of the theater and would be positioned to provide maximum communications coverage. It supports high-bandwidth robust connectivity to command and control elements located over-the-horizon from the littoral/targeted areas.

In the battlefield network, communications requirements are of critical importance between teams or within a team in such a system. A novel concept in mission-oriented communications is **team multicast**. In team multicast the multicast group does not consist of individual members, but rather of teams. For example, a team may be a special task force that is part of a search and rescue mission. The message then must be broadcast to the various teams that are part of the multicast group and to all UVs within each team. For example, a weapon carrying airborne UV may broadcast an image of the target (say, a poison gas plant) to the reconnaissance and sensor teams in front of the formation in order to get a more precise fix on the location of the target. The sensor UV teams that have acquired such information will return the precise location. As another example, suppose N teams with chemical sensors are assessing the "plume" of a chemical spill from different directions. It will be important for each team to broadcast its findings step by step to the other teams using team multicast. In general, team multicast will be commonplace in ad hoc networks designed to support collective tasks, such as those that occur in emergency recovery or in the battlefield.

In the next section we identify the communications needs of teams and then focus on multicast communications.

5.2.2   Opportunistic Ad Hoc Networking in the Urban Grid and on Campus

In this section we describe two sample applications that illustrate the research challenges and the potential power of wireless networks. These are not meant to be a comprehensive list, but are discussed to provide the reader with a meaningful example of wireless networking applications.

Two emerging wireless network scenarios that will soon become part of our daily routines are **vehicle communications** in an urban environment and **campus nomadic networking**. These environments are ripe for benefiting from the technologies discussed in this report. Today, cars connect to the cellular systems mostly for telephony services. The emerging technologies will stimulate an explosion of a new gamut of applications. **Within the car**, short-range wireless communications based on Personal Area Network (**PAN**) technology will be used for monitoring and controlling the vehicle's mechanical components, as well as for connecting the driver's headset to the cellular phone. Another set of innovative applications stems from communications **with other cars** on the road. The potential applications include road safety messages, coordinated navigation, network video games, and other peer-to-peer interactions. These network needs can be efficiently supported by an "opportunistic" **multihop** wireless network among cars which spans the urban road grid and extends to intercity highways. This ad hoc network can alleviate the overload of the fixed wireless infrastructures (3G and hotspot networks). It can also offer an emergency backup in case of a massive fixed infrastructure failure (e.g., terrorist attack, act of war, natural or industrial disaster, etc.). The coupling of a car multihop network, on-board PAN, and cellular wireless infrastructure represents a good example of a **hybrid wireless network** aimed at cost savings, performance improvements, and enhanced resilience to failures.

In the above application the vehicle is a communications hub where the extensive resources of the fixed radio infrastructure and the highly mobile ad hoc radio capabilities meet to provide the necessary services. New networking and radio technologies are needed when operations occur in the "extreme" conditions, namely, extreme mobility (radio and networking), strict delay attributes for safety applications (networking and radio), flexible resource management and reliability (adaptive networks), and extreme throughputs (radios). Extremely flexible radio implementations are needed to realize this goal. Moreover, cross layer adaptation is necessary to explore the trade-offs between transmission rate, reliability, and error control in these environments and to allow the network to gradually adapt as the channel and the application behaviors are better appraised through measurements.

Another interesting scenario is the campus, where the term "campus" here takes the more general meaning of a place where people congregate for various cultural and social (possibly group) activities, thus including amusement park, industrial campus, shopping mall, etc. On a typical campus today, wireless LAN access points in shops, hallways, street crossings, etc. enable nomadic access to the Internet from various portable devices (e.g., laptops, notebooks, PDAs, etc.). However, not all areas of a campus or shopping mall are covered by department/shop wireless LANs. Thus, other wireless media (e.g., GPRS, 1xRTT, 3G) may become useful to fill the gaps. There is a clear opportunity for multiple interfaces or agile radios that can automatically connect to the best available service. The campus will also be an ideal environment where group networking will emerge. For example, on a university campus, students will form small workgroups to exchange files and to share presentations, results, etc. In an Amusement Park, groups of young visitors will interconnect to play network games, etc. Their parents will network to exchange photo shots and video clips. To satisfy this type of close range networking applications, PANs such as BlueTooth and IEEE 802.15 may be brought into the picture. Finally, opportunistic ad hoc networking will become a cost-effective alternative to extend the coverage of access points. Again, as already observed in the vehicular network example, the above "extensions" of the basic infrastructure network model require exactly the technologies recommended in this report, namely, multimode radios, cross layer interaction (to select the best radio interface), and some form of hybrid networking.

These are just simple examples of networked, mobile applications drawn from our everyday lives. These applications, albeit simple, will be immediately enhanced by the new wireless technologies discussed here. There is a wealth of more sophisticated and demanding applications (for example, in the areas of pervasive computing, sensor networks, battlefield, civilian preparedness, disaster recovery, etc.) that will be enabled and spun off by the new radio and network technologies. More examples will be offered in the following sections.

5.3 Design Challenges

As mentioned earlier, ad hoc networks pose a host of new research problems with respect to conventional wireless infrastructure networks. First, we wish to report on some design challenges that cut across the layers. These are mobility, scalability, and cross layer interaction.

5.3.1   Mobility and Scaling

Mobility and reconfiguration are what uniquely distinguish ad hoc networks from other networks. Thus, being able to cope with nodes in motion is an essential requirement. Large scale is also common in ad hoc networks, as battlefield and emergency recovery operations often involve thousands of nodes. The two aspects — mobility and scale — are actually intertwined; anybody can find a workable ad hoc routing solution, say, for 10 nodes, no matter how fast they move; and anybody can find a workable (albeit inefficient) solution (for routing, addressing, service discovery, etc.) for a completely static ad hoc network with, say, 10,000 nodes (just consider the Internet). The problems arise when the 10,000 nodes move at various speeds in various directions over a heterogeneous terrain. In this case, a fixed routing hierarchy such as in the Internet does not work. That is when you have to take out the "big guns" to handle the problem.

Mobility is often viewed as the #1 enemy of the wireless ad hoc network designer. However, mobility, if properly characterized, modeled, predicted, and taken into account, can be of tremendous help in the design of scaleable protocols. In the sequel we offer a few examples where mobility actually helps.

5.3.2   Cross-Layer Interaction

Cross-layer interaction/optimization is a loaded word today, with many different meanings. In ad hoc networks it is a very appropriate way to refer to the fact that it is virtually impossible to design a "universal" protocol layer (routing, MAC, multicast, transport, etc.) and expect that it will function correctly and efficiently in all situations. Predefined protocol layers work reasonably well in the wired Internet. For example, Internet routing, addressing, and Domain Name Server (DNS) protocols work well for both large and small networks. Likewise, the physical and MAC layers of the wired Ethernet are the uncontested reference for all Internet MAC designs. In contrast, in the wireless LAN (the closest relative of the Ethernet), there is convergence not to one, but to a family of standards, from 802.11 to 15 and 16, with each standard addressing different applications. Even within the 802.11 family, a broad range of versions have been defined to address different user scenarios and needs.

In ad hoc network design the importance of tuning the network protocols to the radios and the applications to the network protocols is even more critical, given the extreme range of variability of the systems parameters. Clearly, the routing scheme that works best for a network of a dozen students roaming the Campus may not be suitable for the urban grid with thousand of cars or the battlefield with an extreme range of node speeds and capabilities. Even more important is the concept that in these cases the MAC, routing, and applications must be jointly designed. Moreover, as some parameters (e.g., radio propagation, hostile interference, traffic demands, etc.) may dynamically change, the protocols must be adaptively tuned. Proper tuning requires the exchange of information across layers. For example, in an MIMO radio system the antenna and MAC parameters and possibly routes are dynamically reconfigured based on the state of the channel, which is learned from periodic channel measurements. Thus, interaction between radio channel and protocols is mandatory to achieve an efficient operating point. Video adaptation is another example of cross layer interaction: the video rate stipulated at session initialization cannot be maintained if channel conditions deteriorate. Effective rate adjustment at the source requires careful interplay of end-to-end probing (using, for example, Real-Time Control Protocol [RTCP]) with measurements from the channel and the routing protocol.

5.3.3   The Rest of this Chapter

In the following sections we will address the issues of mobility and scalability in the context of the routing protocol. Routing is fundamental in the management of a mobile ad hoc network. There are a large number of routing protocols in existence today. We will overview the existing routing schemes and then describe some routing solutions recently developed and implemented at UCLA.[47] To set the stage, we will review the Multimedia Intelligent Network of Unattended Mobile Agents (MINUTEMAN) project, which supported much of the routing research at UCLA. We will then introduce the notion of scalable Mobile Backbone Network (MBN) infrastructure and will address the related issues of dynamic Backbone Node (BN) election and stable cluster design. These new protocols will be compared with the state of the art in ad hoc networking.

5.4 Overview of Scalable Ad Hoc Routing Protocols

Due to the facts that bandwidth is scarce in Mobile Ad hoc Networks (MANETs) and the population in a MANET is small compared to the wired Internet, the scalability issue for wireless multihop routing protocols is mostly concerned with the routing message overhead caused by network population size and by mobility. Routing table size is indirectly also a concern because large routing tables imply large control packets and hence high overhead. Routing protocols generally use either distance-vector or link-state routing algorithms.[21] Both types find the shortest paths to destinations. In distance-vector (DV) routing, a vector containing the cost (e.g., hop distance) and path (next hop) to all the destinations is kept and exchanged at each node. DV protocols are generally known to suffer from slow route convergence and a tendency to create loops in mobile environments. The link-state (LS) routing algorithm overcomes the problem by maintaining global network topology information at each router through periodical flooding of link information about its neighbors. Mobility entails frequent flooding. Unfortunately, this LS advertisement scheme generates larger routing control overhead than DV. In a network with population N, LS updating generates routing overhead on the order of $O(NxN)$. In large networks, the transmission of routing information will ultimately consume most of the bandwidth. Thus, reducing routing control overhead becomes a key issue for routing scalability.

In some application domains (e.g., digitized battlefield), scalability is realized by designing a hierarchical architecture with physically distinct layers (e.g., point-to-point wireless backbone).[22] However, such a physical hierarchy is not cost effective for many other applications (e.g., sensor networks). Thus, it is important to find scalable solutions using a homogeneous network (as opposed to a network with multiple physical layers).

The scalability is more challenging in the presence of both large numbers and mobility. If nodes are stationary, the large population can be effectively handled with conventional hierarchical routing. In contrast, when nodes move, hierarchical partitioning must be continuously updated. As for mobility management, the Mobile IP solution works well if mobility is limited to a few users and there is a fixed infrastructure that hosts the *home agent*, namely, the node in the fixed network that keeps track of the mobile node.[29] When all nodes move (including the home agent), such a strategy cannot be applied.

A considerable body of literature has addressed research on routing and architecture of ad hoc networks. Relating to the problem described above, we present a survey with a focus on solutions toward scalability in large populations that are able to handle mobility. Classification according to routing strategy, that is, proactive (or table-driven) and reactive (or on-demand), has been used in several articles.[18,23–26] Here, we provide a slightly different classification that is based on the routing protocol structure. Different structures affect the design and operation of the routing protocols; they also determine the performance with regard to scalability. Reviews and performance comparisons of ad hoc routing protocols have also been presented in previous publications.[23,24,26,27,29,30] While some overlap with previous work is inevitable in order to preserve the integrity of our presentation, our choice of protocols includes recent examples that reveal unique features in term of scalability.

In the sequel we review key routing protocols in three categories (Figure 5.2):

- Flat routing schemes, which are further classified into two classes: proactive and reactive, according to their design philosophy

- Hierarchical routing

- Geographic position assisted routing

Flat routing approaches adopt a flat addressing scheme. Each node participating in routing plays an equal role. In contrast, hierarchical routing requires a hierarchical address and usually assigns different

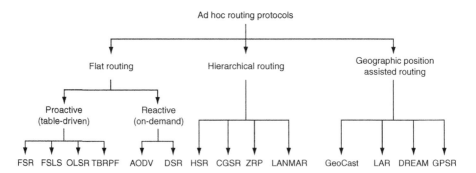

Figure 5.2 Classification of ad hoc routing protocols. (Note: The acronyms in the figure are explained throughout the chapter.)

roles to different nodes in the hierarchy. Routing with geographic assistance requires each node to be equipped with the Global Position-ing System (GPS). This requirement is quite realistic today since such devices are inexpensive and can provide reasonable precision.

5.4.1   Routing in a Flat Network Address Structure

The protocols we review here fall into two categories: proactive (table-driven) and reactive (on-demand) routing. Many proactive protocols stem from conventional LS routing. On-demand routing, on the other hand, is a new emerging routing philosophy in the ad hoc area. It differs from conventional routing protocols in that no routing activi-ties and no permanent routing information are maintained at network nodes if there is no communication, thus providing a scalable routing solution to large populations.

5.4.1.1   Proactive Routing Protocols

Proactive routing protocols share a common feature, that is, back-ground routing information exchange regardless of communication requests. The protocols have many desirable properties, especially for applications including real-time communications and Quality of Ser-vice (QoS) guarantees, such as low latency route access and alternate QoS path support and monitoring. Many proactive routing protocols have been proposed for efficiency and scalability.

Fisheye State Routing

Fisheye State Routing (FSR) described in Pei, Gerla, and Chen[20] and Iwata et al.[3] is a simple, efficient LS type routing protocol that main-tains a topology map at each node and propagates LS updates. The main differences between FSR and conventional LS protocols are the ways in which routing information is disseminated. First, FSR exchanges the entire LS information only with neighbors instead of flooding it over the network. The LS table is kept up to date based on the information received from neighbors. Second, the LS exchange is periodical instead of event triggered, which avoids frequent LS updates caused by link breaks in an environment with unreliable wire-less links and mobility. Moreover, the periodical broadcasts of the LS information are conducted in different frequencies for different entries depending on their hop distances to the current node. Entries corresponding to faraway (outside a predefined *scope*) destinations are propagated with a lower frequency than those corresponding to nearby destinations. As a result, a considerable fraction of entries are suppressed from LS exchange packets. FSR produces accurate distance and path information about the immediate neighborhood of a node and imprecise knowledge of the best path to a distant destination. However, this imprecision is compensated by the fact that the route on which the packet travels becomes progressively more accurate as the packet approaches its destination. Similar work is also presented in Fuzzy Sighted Link State (FSLS) routing,[5–25] FSLS includes an optimal algorithm called Hazy Sighted Link State (HSLS), which sends a link state update (LSU) every $2k^* T$ to a scope of $2k$, where k is hop distance and T is the minimum LSU transmission period. Thus, both FSR and FSLS achieve potential scalability by limiting the scope of LSU dissemination in space and over time. Theoretical analysis on routing overhead and optimization for this type of "myopic" routing can be found in Santivanez, Ramanathan, and Stavrakakis.[25]

Optimized Link State Routing Protocol

Optimized Link State Routing (OLSR) protocol[32] is an LS protocol. It periodically exchanges topology information with other nodes in the network. The protocol uses *multipoint relays (MPRs)*[5–33] to reduce the number of "superfluous" broadcast packet retransmissions and the size of the LSU packets, leading to efficient flooding of control messages in the network.

A node, say, node A, periodically broadcasts HELLO messages to all immediate neighbors to exchange neighborhood information (i.e., list of neighbors) and to compute the MPR set. From neighbor lists, node A figures out the nodes that are two hops away and computes the minimum set of one-hop relay points required to reach the two-hop neighbors. Such a set is the MPR set. Figure 5.3 illustrates the MPR set of node A. The optimum (minimum size) MPR computation is of combinstorial complexity (i.e., Non-Polynominal complete). Efficient heuristics are used. Each node informs its neighbors about its MPR set in the HELLO message. Upon receiving such a HELLO, each node records the nodes (called MPR *selectors*) that select it as one of their MPRs. In routing information dissemination, OLSR differs from pure LS protocols in two aspects. First, by construction, only the MPR nodes of A need to forward the LSUs issued by A. Second, the LSU of node A is reduced in size since it includes only the neighbors that select node A as one of their MPR nodes. In this way, partial topology information

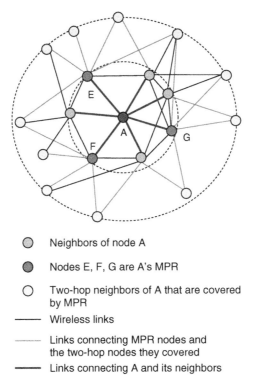

○ Neighbors of node A

● Nodes E, F, G are A's MPR

○ Two-hop neighbors of A that are covered by MPR

—— Wireless links

----- Links connecting MPR nodes and the two-hop nodes they covered

—— Links connecting A and its neighbors

Figure 5.3 OLSR: an illustration of multipoint relays.

is propagated, that is, say, node A can be reached only from its MPR selectors. OLSR computes the shortest path to an arbitrary destination using the topology map consisting of all of its neighbors and of the MPRs of all other nodes. OLSR is particularly suited for dense networks. When the network is sparse, every neighbor of a node becomes an MPR. The OLSR then reduces to a pure LS protocol.

Topology Broadcast Based on Reverse Path Forwarding

Topology Broadcast Based on Reverse Path Forwarding (TBRPF)[34,35] is also an LS protocol. It consists of two separate modules: the Topology and Neighbor Discovery (TND) module and the routing module. TND is performed through periodical "differential" HELLO messages that report only the changes (up or lost) of neighbors. The TBRPF routing module operates based on partial topology information obtained through both periodic and differential topology updates. Operation in full topology is provided as an option by including additional topology information in updates.

TBRPF works as follows. Assume node S is the source of update messages. Every node i in the network chooses its next hop (say, node p) on the minimum-hop path toward S as its parent with respect to node S. Instead of flooding to the entire net, TBRPF only propagates LS updates in the reverse direction on the spanning tree formed by the minimum-hop paths from all nodes to the source of the updates, that is, node i only accepts topology updates originated at node S from parent node p and then forwards them to the children pertaining to S. Furthermore, only the links that will result in changes to i's source tree are included in the updates. Thus, a smaller subset of the source tree is propagated. The leaves of the broadcast tree do not forward updates. Each node can also include the entire source tree in the updates for full topology operation. The topology updates are broadcast periodically and differentially. The differential updates are issued more frequently to quickly propagate link changes (additions and deletions). Thus, TBRPF adapts to topology change faster, generates less routing overhead, and uses a smaller topology update packet size than pure LS protocols.

5.4.1.2 Reactive (on-demand) Routing Protocols

Reactive routing (also referred to as "on-demand routing") is a popular routing category for wireless ad hoc routing. The design follows the idea that each node tries to reduce routing overhead by only sending routing packets when a communication is awaiting. Examples include Ad hoc On-demand Distance-Vector routing (AODV),[13] Associativity-Based Routing (ABR),[36] Dynamic Source Routing (DSR),[14] Lightweight Mobile Routing (LMR),[37] and Temporally Ordered Routing Algorithms (TORA).[38] Among the many proposed protocols, AODV and DSR have been extensively evaluated in the MANET literature and are being considered by the Internet Engineering Task Force (IETF) MANET Working Group as the leading candidates for standardization. They are described briefly here to demonstrate the reactive routing mechanism. Interested readers are referred to Das and coworkers[10,23] and Haas and Pearlman[26] for performance evaluations.

Reactive algorithms typically have a route discovery phase. Query packets are flooded into the network by the sources in search of a path. The phase completes when a route is found or all the possible outgoing paths from the source are searched. There are different approaches for discovering routes in reactive algorithms. In AODV, on receiving a query, the transit nodes "learn" the path to the source (called *backward learning*) and enter the route in the forwarding table. The intended destination eventually receives the query and can thus respond using the path traced by the query. This permits the establishment of a full duplex path. To reduce new path search overhead, the query packet is dropped during flooding if it encounters a node which already has a route to the destination. After the path has been established, it is maintained as long as the source uses it. A link failure will be reported to the source recursively through the intermediate nodes. This, in turn, will trigger another query-response procedure in order to find a new route.

An alternate scheme for tracing paths in a reactive fashion is DSR. DSR uses *source routing*, that is, a source indicates in a data packet's header the sequence of intermediate nodes on the routing path. In DSR, the query packet copies in its header the IDs of the intermediate nodes it has traversed. The destination then retrieves the entire path from the query packet and uses it (via source routing) to respond to the source, providing the source with the path at the same time. Data packets carry the source route in the packet headers. A DSR node aggressively caches the routes it has leaned so far to minimize the cost incurred by the route discovery. Source routing enables DSR nodes to keep multiple routes to a destination. When link breakage is detected (through *passive acknowledgments*), route reconstruction can be delayed if the source can use another valid route directly. If no such alternate routes exist, a new search for a route must be reinvoked. The path included in the packet header makes the detection of loops very easy.

To reduce the route search overhead, both protocols provide optimizations by taking advantage of existing route information at intermediate nodes. *Promiscuous listening* (overhearing neighbor propagation) used by DSR helps nodes learn as many route updates as they can without actually participating in routing. *Expanding ring search* (controlled by the *time-to-live* field of route request packets) used by AODV limits the search area for a previous discovered destination using the prior hop distance.

5.4.1.3   Comparisons of Flat Address Routing Protocols
Key characteristics of the protocols discussed in this section are summarized in Table 5.1. In the table, N denotes the number of nodes in the network, and e denotes the number of communication pairs. Storage complexity measures the order of the table size used by the protocols. Communication complexity gives the number of messages needed to perform an operation when an update occurs.

Table 5.1 Characteristics of flat routing protocols.

	FSR	OLSR	TBRPF	AODV	DSR
Routing philosophy	Proactive	Proactive	Proactive	On-demand	On-demand
Routing metric	Shortest path	Shortest path	Shortest path	Shortest path	Shortest path
Frequency of updates	Periodically	Periodically	Periodically, as needed (link changes)	As needed (data traffic)	As needed (data traffic)
Use sequence numbers	Yes	Yes	Yes (HELLO)	Yes	No
Loop-free	Yes	Yes	Yes	Yes	Yes
Worst case exists	No	Yes (pure LS)	No	Yes (full flooding)	Yes (full flooding)
Multiple paths	Yes	No	No	No	Yes
Storage complexity	$O(N)$	$O(N)$	$O(N)$	$O(e)$	$O(e)$
Comm. complexity	$O(N)$	$O(N)$	$O(N)$	$O(2N)$	$O(2N)$

The proactive protocols adopt different ways toward scalability. FSR introduces the notion of multilevel fisheye scope to reduce routing update overhead through reducing the routing packet sizes and update frequency. FSLS/HSLS further drives this limited dissemination approach to an optimal point. OLSR produces less control overhead than FSR because it forces the propagation of LSUs only at MPR nodes, leading to fewer nodes participating in LSU forwarding. Similarly, TBRPF reduces the LSUs forwarding at leaf nodes of each source tree and disseminates differential updates. It also generates smaller HELLO messages than OLSR. Both OLSR and TBRPF achieve more efficiency than classic LS algorithms when networks are dense, that is, OLSR obtains a larger compression ratio of number of MPRs over number of neighbors, and TBRPF trims more leaf nodes from propagation. The multilevel scope reduction from FSR and FSLS, however, will not reduce propagation frequency when a network grows dense. In contrast, the scope reduction works well when a network grows in diameter (in terms of hop distance). Multiple scopes can effectively reduce the update frequency for nodes many hops away. However, all four protocols require nodes to maintain routing tables containing entries for all the nodes in the network (storage complexity $O(N)$). This is acceptable if the user population is small. As the number of mobile hosts increases, so does the overhead. This affects the scalability of the protocols in large networks.

Reactive protocols "react" only to communication needs. The routing overhead thus relates to the discovery and maintenance of the routes in use. With light traffic (directed to a few destinations) and low mobility, reactive protocols scale well to large populations (low bandwidth and storage overhead). However, for heavy traffic with a large number of destinations, more sources will search for destinations. Also, as mobility increases, the discovered routes may soon break down, requiring repeated route discoveries. Route caching also becomes ineffective with high mobility. Since flooding is used for query dissemination and route maintenance, routing control overhead tends to grow very high.[10] Longer delays are also expected in large mobile networks. In addition, DSR generates larger routing and data packets due to the path information stored in the header. In large networks where long paths prevail, the DSR header becomes a major contribution to overhead.

In terms of scattered traffic pattern and high mobility, proactive protocols produce higher routing efficiency than reactive protocols. In the former, the routes to all the destinations are known in advance. Fresh route information is maintained periodically. No additional routing overhead needs to be generated for finding a new destination or route. The flip side is that proactive protocols constantly consume bandwidth and energy due to the periodic updates. This property makes proactive schemes undesirable for some resource critical applications (e.g., sensor networks).

For AODV and DSR, since a route has to be entirely discovered prior to the actual data packet transmission, the initial search latency may degrade the performance of interactive applications (e.g., distributed database queries). In contrast, FSR, OLSR, and TBRPF avoid the extra work of "finding" the destination by retaining a routing entry for each destination all the time, thus providing low single-packet transmission latency. Proactive schemes such as FSR, OLSR, and TBRPF can be easily extended to support QoS by including bandwidth and channel quality information in LS entries. Thus, the quality of the path (e.g., bandwidth, delay) is known prior to call setup. For AODV and DSR, the quality of the path is not known a priori. It can be discovered only while setting up the path and must be monitored by all intermediate nodes during the session, at the cost of additional latency and overhead penalty.

5.4.2 Hierarchical Routing Protocols

Typically, when wireless network size increases (beyond certain thresholds), current "flat" routing schemes become infeasible because of link and processing overhead. One way to solve this problem, and to produce scalable and efficient solutions, is hierarchical routing. An example of hierarchical routing is the Internet hierarchy, which has been practiced in the wired network for a long time. Wireless hierarchical routing is based on the idea of organizing nodes in groups and then assigning nodes different functionalities inside and outside a group. Both routing table size and update packet size are reduced by including in them only part of the network (instead of the whole); thus, control overhead is reduced. The most popular way of building hierarchy is to group nodes geographically close to each other into explicit clusters. Each cluster has a leading node (*clusterhead*) to communicate to other nodes on behalf of the cluster. An alternate way is to have implicit hierarchy. In this way, each node has a local scope. Different routing strategies are used inside and outside the scope. Communications pass across overlapping scopes. A more efficient overall routing performance can be achieved through this flexibility. Since mobile nodes have only a single omnidirectional radio for wireless communications, this type of hierarchical organization will be referred to as *logical hierarchy* to distinguish it from the physically hierarchical network structure.

5.4.2.1 Clusterhead-Gateway Switch Routing

Clusterhead-Gateway Switch Routing (CGSR)[38] is typical of cluster-based hierarchical routing. A stable clustering algorithm, *Least Clusterhead Change* (LCC), is used to partition the whole network into clusters, and a *clusterhead* is elected in each cluster. A mobile node that belongs to two or more clusters is a *gateway* connecting the clusters. Data packets are routed through paths having a format of

"Clusterhead–Gateway–Clusterhead–Gateway–Clusterhead..." and so on between any source and destination pairs.

CGSR is a DV routing algorithm. Two tables, a cluster member table and a DV routing table, are maintained at each mobile node. The cluster member table records the clusterhead for each node and is broadcast periodically. A node will update its member table on receiving such a packet. The routing table only maintains one entry for each cluster recording the path to its clusterhead, no matter how many members it has. To route a data packet, the current node first looks up the clusterhead of the destination node from the cluster member table. Then it consults its routing table to find the next hop to that destination cluster and routes the packet toward the destination clusterhead. The destination clusterhead will finally route the packet to the destination node, which is a member of it and can be directly reached. This procedure is demonstrated in Figure 5.4.

The major advantage of CGSR is that it can greatly reduce the routing table size compared to DV protocols. Only one entry is needed for all nodes in the same cluster. Thus, the broadcast packet size of the routing table is reduced. These features make a DV routing scale to large network size. Although an additional cluster member table is required at each node, its size is only decided by the number of clusters in the network. The drawback of CGSR is the difficulty of maintaining the cluster structure in a mobile environment. The LCC clustering algorithm introduces additional overhead and complexity in the formation and maintenance of clusters.

5.4.2.2 Hierarchical State Routing
Hierarchical State Routing (HSR) (see Figure 5.5) is a multilevel clustering-based LS routing protocol.[40] HSR maintains a logical

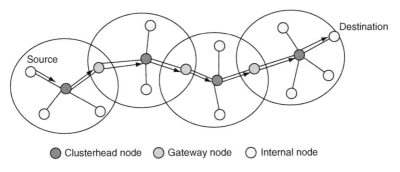

Figure 5.4 CGSR routing: showing a data path from source to destination.

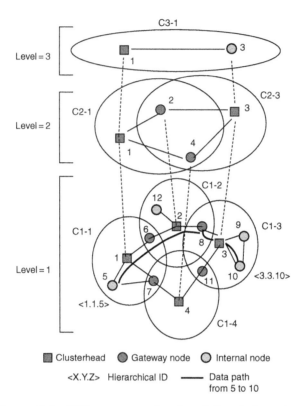

Figure 5.5 HSR: an example of multilevel clustering.

hierarchical topology by using the clustering scheme recursively. Nodes at the same logical level are grouped into clusters. The elected clusterheads at the lower level become members of the next higher level. These new members, in turn, organize themselves in clusters, and so on. The goal of clustering is to reduce routing overhead (i.e., routing table storage, processing, and transmission) at each level. An example of a three-level hierarchical structure is demonstrated in Figure 5.5. Generally, there are three kinds of nodes in a cluster: clusterheads (e.g., nodes 1, 2, 3, and 4), gateways (e.g., nodes 6, 7, 8, and 11), and internal nodes (e.g., nodes 5, 9, 10, and 12). A clusterhead acts as a local coordinator for transmissions within the cluster.

HSR is based on LS routing. At the first level of clustering (also the physical level), each node monitors the state of the link to each neighbor (i.e., link up/down and possibly QoS parameters, e.g., bandwidth) and broadcasts it within the cluster. The clusterhead summarizes LS information within its cluster and propagates it to the neighbor clusterheads (via the gateways). The knowledge of connectivity between neighbor clusterheads leads to the formation of level 2 clusters. For example, as shown in Figure 5.5, neighbor clusterheads 1 and 2 become members of the level 2 cluster C2. LS entries at level 2 nodes contain the "virtual" links in C2. A virtual link between neighbor nodes 1 and 2 consists of the level 1 path from clusterhead 1 to clusterhead 2 through gateway 6. The virtual link can be viewed as a "tunnel" implemented through lower level nodes. Applying the aforementioned clustering procedure recursively, new clusterheads are elected at each level and become members of the higher level cluster. If QoS parameters are required, the clusterheads will summarize the information from the level they belong to and carry it into the higher level. After obtaining the LS information at one level, each virtual node floods it down to nodes of the lower level clusters. As a result, each physical node has "hierarchical" topology information through the hierarchical address of each node (described below), as opposed to a full topology view as in flat LS schemes.

The hierarchy so developed requires a new address for each node, the hierarchical address. The node labels from 1 to 12 shown in Figure 5.5 (at level = 1) are physical (e.g., MAC layer) addresses. They are hardwired and unique to each node. In HSR, the *Hierarchical ID* (HID) of a node is defined as the sequence of MAC addresses of the nodes on the path that starts from the top hierarchy to the node itself. For example, in Figure 5.5 the top of the hierarchy is node 1, and the hierarchical address of node 5, HID(5), is the path from node 1 to node 5, i.e., $<1, 1, 5>$. Likewise, HID $(10) = <3, 3, 10>$. The advantage of this hierarchical address scheme is that each node can dynamically and locally update its own HID upon receiving the routing updates from the nodes higher up in the hierarchy. The hierarchical address is sufficient to deliver a packet to its destination from anywhere in the network using HSR tables. Gateway nodes can communicate with multiple clusterheads and thus can be reached from the top hierarchy via multiple paths. Consequently, a gateway has multiple hierarchical addresses, similar to a router in the wired Internet, equipped with multiple subnet addresses. These benefits come at the cost of longer (hierarchical) addresses and frequent updates of the cluster hierarchy and the hierarchical addresses as nodes move. In principle, a continuously changing hierarchical address makes it difficult to locate and keep track of nodes.

5.4.2.3 Zone Routing Protocol

The Zone Routing Protocol (ZRP)[26] is a hybrid routing protocol that combines both proactive and on-demand routing strategies and benefits from advantages of both types. The basic idea is that each node has a predefined *zone* centered at itself in terms of number of hops. For nodes within the zone, it uses proactive routing protocols to maintain routing information. For those nodes outside of its zone, it does not maintain routing information in a permanent base. Instead, on-demand routing strategy is adopted when interzone connections are required.

The ZRP protocol consists of three components. Within the zone, proactive *Intrazone Routing Protocol (IARP)* is used to maintain routing information. IARP can be any LS routing or DV routing depending on the implementation. For nodes outside the zone, reactive *Interzone Routing Protocol (IERP)* is performed. IERP uses the *route query* (RREQ)/*route reply* (RREP) packets to discover a route in a way similar to typical on-demand routing protocols. IARP always provides a route to nodes within a node's zone. When the intended destination is not known at a node (i.e., not in its IARP routing table), that node must be outside of its zone. Thus, an RREQ packet is broadcast via the nodes on the border of the zone. Such an RREQ broadcast is called *Bordercast Resolution Protocol* (BRP). Route queries are only broadcast from one node's border nodes to other border nodes until one node knows the exact path to the destination node (i.e., the destination is within its zone). The hybrid proactive/reactive scheme limits the proactive overhead to only the size of the zone and the reactive search overhead to only selected border nodes. However, potential inefficiency may occur when flooding of the RREQ packets goes through the entire network.

5.4.2.4 Landmark Ad Hoc Routing Protocol

Landmark ad hoc routing protocol (LANMAR)[17,18] is designed for an ad hoc network that exhibits group mobility. Namely, one can identify logical subnets in which the members have a commonality of interests and are likely to move as a group (e.g., a brigade or tank battalion in the battlefield). LANMAR uses an IP-like address consisting of a group ID (or subnet ID) and a host ID: *<GroupID, HostID>*. LANMAR uses the notion of *landmarks* to keep track of such logical groups. Each logical group has one dynamically elected node serving as a *landmark*. A global DV mechanism (e.g., Destination-Sequenced Distance-Vector [DSDV][15]) propagates the routing information about all the landmarks in the entire network. Furthermore, LANMAR works in symbiosis with a local scope routing scheme. The local routing scheme can use the flat proactive protocols mentioned previously (e.g., FSR). FSR maintains detailed routing information for nodes within a given scope D (i.e., FSR updates propagate only up to hop distance D). As a result, each node has detailed topology information about nodes within its local scope and has a distance and routing vector to all landmarks. When a node needs to relay a packet to a destination within its scope, it uses the FSR routing tables directly. Otherwise, the packet will be routed toward the landmark corresponding to the destination's logical subnet, which is read from the logical address carried in the packet header. When the packet arrives within the scope of the destination, it is routed using local tables (that contain the destination), possibly without going through the landmark.

LANMAR reduces both routing table size and control overhead effectively through the truncated local routing table and "summarized" routing information for remote groups of nodes. In general, by adopting different local routing schemes,[30] LANMAR provides a flexible routing framework for scalable routing while still preserving the benefits introduced by the associated local scope routing scheme.

Comparisons of Hierarchical Routing Protocols

Table 5.2 summarizes the features of the four hierarchical routing protocols. In the table, N is the total number of mobile nodes in the network; M is the average number of nodes in a cluster; and L is the average number of nodes in a node's local scope, which is used by both ZRP and LANMAR and is given here an identical scope size (r hops). The difference between M and L is that M usually only includes one-hop nodes, while L includes nodes up to r hops. Also, in Table 5.2, H is the number of hierarchical levels of HSR, and G is the number of logical groups in LANMAR. The number of communication pairs is denoted as e. The storage and communication complexity have the same definitions as given in an earlier section.

The explicit hierarchical protocols CGSR and HSR force a path to go through some critical nodes like clusterheads and gateways, leading to possibly suboptimal paths. The two implicitly hierarchical protocols ZRP and LANMAR use a shortest path algorithm at each node. However, LANMAR guarantees shortest paths only when destinations are within the scope. For remote nodes, though data packets are first routed toward remote landmarks through shortest paths, extra hops

Table 5.2   Characteristics of hierarchical routing protocols.

	CGSR	HSR	ZRP	LANMAR
Hierarchy	Explicit two levels	Explicit multiple levels	Implicit two levels	Implicit two levels
Routing philosophy	Proactive, DV	Proactive, LS	Hybrid, DV and LS	Proactive, DV and LS
Loop-free	Yes	Yes	Yes	Yes
Routing metric	Via critical nodes	Via critical nodes	Local shortest path	Local shortest path
Critical nodes	Yes (clusterhead)	Yes (clusterhead)	No	Yes (landmark)
Storage complexity	$O(N/M)$	$O(M^*H)$	$O(L) + O(e)$	$O(L) + O(G)$
Comm. complexity	$O(N)$	$O(M^*H)$	$O(N)$	$O(N)$

may be traveled before a destination is hit. Similarly, ZRP does not provide an overall optimized shortest path if the destination has to be found through IERP.

CGSR maintains two tables at each node: a cluster member table and a routing table. The routing table contains one route to each cluster (actually clusterhead). Its storage complexity is $O(N/M)$. For the cluster member table, again, only one entry is needed for each cluster. Thus, the storage complexity of CGSR is $O(N/M)$. In HSR, nodes at different levels have different storage requirements. The worst case occurs at the top level. The top-level nodes have to maintain a routing table of its clusters at each level. Thus, its storage complexity is $O(M \times H)$. ZRP has separate tables for IARP and IERP. IARP is proactive and its storage complexity is $O(L)$. IERP is reactive routing; thus, the table size depends on the traffic pattern, leading to storage on the order of $O(L) + O(e)$. In LANMAR routing, each node also keeps two routing tables. One is a local routing table keeping track of all nodes in the scope. The other is a DV routing table maintaining paths to all landmarks. Thus, its storage complexity is $O(L) + O(G)$. Usually, the number of groups (G) is small (comparing to network size N). For example, for a simple network with equal partitions, when group size is 25 nodes, a 100-node network has 4 groups, and 1000-node network generates 40 groups.

The communication complexity of CGSR is $O(N)$, since the routing table and cluster member table have to be propagated throughout the whole network. Link updates in HSR are propagated along the hierarchical tree. In the worst case, if the top-level clusterhead is changed, corresponding worst case communication complexity is $O(M \times H)$. The worst case in ZRP occurs when a link change requires rediscovery of a new route over the entire network; thus, the communication complexity is $O(N)$. In LANMAR, though the local proactive protocol has a communication complexity on the order of $O(L)$, the total complexity is still $O(N)$ as the landmark DVs have to be propagated throughout the whole network.

The comparisons of the storage and communication complexities show that hierarchical routing protocols maintain smaller routing tables compared to flat proactive routing protocols. Even though the basic protocols have an equivalent communication complexity as in flat routing, routing overhead is greatly reduced because a smaller message size is used. For example in HSR, the storage $O(M \times H)$ can be expressed as $O(M \times \log N / \log M)$ (because the total number of nodes N can be expressed as $O(MH)$), and the routing overhead is $O((M \times \log N / \log M)2)$. In LANMAR, the routing overhead is $O((L + G) \times N)$. Both are smaller than $O(N2)$ in flat LS routing. The reduction in overhead greatly improves hierarchical routing protocol scalability to large network sizes.

However, in the face of mobility, explicit cluster-based hierarchical protocols will induce additional overhead in order to maintain the hierarchical structure. HSR further requires complex management for HID registrations and translations.[41] This will not be the case for the "implicitly hierarchical" ZRP and LANMAR.

Both ZRP and LANMAR use proactive routing for local operations. However, they differ in outside scope routing. ZRP adopts an on-demand scheme, and LANMAR uses a proactive scheme. Thus, when network size increases and destinations are more likely to be outside the local scope, ZRP's behavior becomes similar to on-demand routing with unpredictable large overhead, while LANMAR has the advantage that the landmark DV is small and grows slowly. LANMAR greatly improves routing scalability to large MANETs. The main limitation of LANMAR is the assumption of group mobility.

5.4.3   Geographic Position Information Assisted Routing

The advances in the development of GPS nowadays make it possible to provide location information with a precision within a few meters. It also provides universal timing. While location information can be used for directional routing in distributed ad hoc systems, the universal clock can provide global synchronizing among GPS-equipped nodes. Research has shown that geographical location information can improve routing performance in ad hoc networks. Additional care must be taken into account in a mobile environment, because locations may not be accurate by the time the information is used. All the protocols surveyed below assume that the nodes know their positions.

5.4.3.1   Geographic Addressing and Routing

Geographic Addressing and Routing (GeoCast)[42] allows messages to be sent to all nodes in a specific geographical area using geographic information instead of logical node addresses. A geographic destination address is expressed in three ways: point, circle (with center point and radius), and polygon (a list of points, e.g., $P(1)$, $P(2), \ldots$, $P(n-1)$, $P(n)$, $P(1)$). A point is represented by geographic coordinates (latitude and longitude). When the destination of a message is a polygon or circle, every node within the geographic region of the polygon/circle will receive the message. A geographic router (*GeoRouter*) calculates its *service area* (geographic area it serves) as the union of the geographic areas covered by the networks attached to it (Figure 5.6). This service area is approximated by a single closed polygon. GeoRouters exchange service area polygons to build routing tables. This approach builds a hierarchical structure (possibly wireless) consisting of GeoRouters. The end users can move freely about the network.

Data communication starts from a computer host capable of receiving and sending geographic messages (GeoHost). Data packets are then sent to the local GeoNode (residing in each subnet), which is responsible for forwarding the packets to the local GeoRouter. A GeoRouter first checks whether its service area intersects the destination polygon. As long as a part of the destination area is not covered, the GeoRouter sends a copy of the packet to its parent router for further routing beyond its own service area. Then it checks the service area of its child routers for a possible intersection. All the child routers intersecting the target area are sent a copy of the packet. When a router's service area

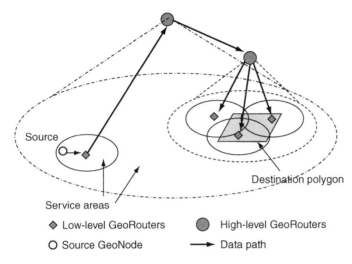

Figure 5.6 An example of GeoCast.

falls within the target area, the router picks up the packet and forwards it to the GeoNodes attached to it. Figure 5.6 illustrates the procedure of routing over GeoRouters.

As GeoCast is designed for group reception, multicast groups for receiving geographic messages are maintained at the GeoNodes. The incoming geographic messages are stored for a lifetime (determined by the sender), and during that time, they are multicast periodically through an assigned multicast address. Clients at GeoHosts tune into the appropriate multicast address to receive the messages.

5.4.3.2 Location-Aided Routing

The Location-Aided Routing (LAR) protocol presented in Ko and Vaidya[43] is an on-demand protocol based on source routing. The protocol utilizes location information to limit the area for discovering a new route to a smaller *request zone*. As a consequence, the number of route request messages is reduced.

The operation of LAR is similar to DSR.[14] Using location information, LAR performs the route discovery through *limited flooding* (i.e., floods the requests to a request zone). Only nodes in the request zone will forward route requests. LAR provides two schemes to determine the request zone.

- **Scheme 1**: The source estimates a circular area (*expected zone*) in which the destination is expected to be found at the current time.

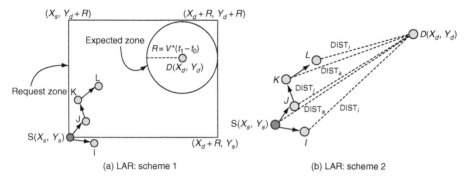

(a) LAR: scheme 1

(b) LAR: scheme 2

Figure 5.7 LAR: limited flooding of route request: (a) Scheme 1, expected zone; (b) Scheme 2, closer distances.

The position and size of the circle are calculated based on the knowledge of the previous destination location, the time instant associated with the previous location record, and the average moving speed of the destination. The smallest rectangular region that includes the expected zone and the source is the request zone (Figure 5.7(a)). The coordinates of the four corners of the zone are attached to a route request by the source. During the route request flood, only nodes inside the request zone forward the request message.

- **Scheme 2**: The source calculates the distance to the destination based on the destination location known to it. This distance, along with the destination location, is included in a route request message and sent to neighbors. When a node receives the request, it calculates its distance to the destination. A node will relay a request message only if its distance to the destination is less than or equal to the distance included in the request message. For example, in Figure 5.7(b), nodes I and J will forward the requests from S. Before a node relays the request, it updates the distance field in the message with its own distance to the destination.

5.4.3.3 Distance Routing Effect Algorithm for Mobility

Distance Routing Effect Algorithm for Mobility (DREAM)[44] is a proactive routing protocol using location information. It provides distributed, loop-free, multipath routing and is able to adapt to mobility. It minimizes the routing overhead by using two new principles for the routing update frequency and message lifetime. The principles are *distance effect* and *mobility rate*. With the distance effect, the greater the distance separating two nodes, the slower they appear to be moving with respect to each other. With the mobility rate, the faster a node moves, the more frequently it needs to advertise its new location. Using the location information obtained from GPS, each node can realize the two principles in routing.

In DREAM, each node maintains a *location table* (LT). The table records locations of all the nodes. Each node periodically broadcasts control packets to inform all other nodes of its location. The distance effect is realized by sending more frequently to nodes that are more closely positioned. In addition, the frequency of sending a control packet is adjusted based on its moving speed.

With the location information stored at routing tables, data packets are partially flooded to nodes in the direction of the destination. The source first calculates the direction toward the destination, and then it selects a set of one-hop neighbors that are located in the direction. If this set is empty, the data is flooded to the entire network. Otherwise, the set is enclosed in the data header and transmitted with the data. Only nodes specified in the header are qualified to receive and process the data packet. They repeat the same procedure by selecting their own set of one-hop neighbors, updating the data header, and sending the packet out. If the selected set is empty, the data packet is dropped. When the destination receives the data, it responds with an Acknowledgment (ACK) to the source in a similar way. However, the destination will not issue an ACK if the data is received via flooding. If the source does not receive an ACK for data sent through a designated set of nodes, it retransmits the data again by pure flooding.

5.4.3.4   Greedy Perimeter Stateless Routing

Greedy Perimeter Stateless Routing (GPSR)[45] is a routing protocol that uses only neighbor location information in forwarding data packets. It requires only a small amount of per-node routing state, has low routing message complexity, and works best for dense wireless networks. In GPSR, beacon messages are periodically broadcast at each node to inform its neighbors of its position, which results in minimized one-hop-only topology information at each node. To further reduce the beacon overhead, the position information is piggybacked in all the data packets a node sends. GPSR assumes that sources can determine through separate means the location of destinations and can include such locations in the data packet header. A node makes forwarding decisions based on the relative position of destination and neighbors.

GPSR uses two data forwarding schemes: *greedy forwarding* and *perimeter forwarding*. The former is the primary forwarding strategy, while the latter is used in regions where the primary one fails. Greedy forwarding works this way: when a node receives a packet with the destination's location, it chooses from its neighbors the node that is geographically closest to the destination and then forwards the data packet to it. This local optimal choice repeats at each intermediate node until the destination is reached. When a packet reaches a dead end (i.e., a node whose neighbors are all farther away from the destination than itself), perimeter forwarding is performed.

Before performing the perimeter forwarding, the forwarding node needs to calculate a *relative neighborhood graph* (RNG), that is, for all the neighbor nodes, the following inequality holds:

$$\forall w \neq u, v : d(u, v) \leq max[d(u, w), d(v, w)], \tag{5.1}$$

where u, v, and w are nodes, and $d(u,v)$ is the distance of edge (u,v). A distributed algorithm of removing edges violating inequality 1 from the original neighbor list yields a network without crossing links and retaining connectivity.

Perimeter forwarding traverses the RNG using the right-hand rule hop by hop along the perimeter of the region. During perimeter forwarding, if the packet reaches a location that is closer to the destination than the position where the previous greedy forwarding of the packet failed, the greedy process is resumed. Possible loops during perimeter forwarding occur when the destination is not reachable. These will be detected, and the packets will be dropped. In the worst case, GPSR will possibly generate a very long path before a loop is detected.

Comparisons of Geographic Position Assisted Routing

With the knowledge of node locations, routing can be more effective and scalable in the realm of routing philosophy at the cost of the overhead incurred by exchanging coordinates. Key characteristics and properties of the protocols are summarized in Table 5.3. The same notations used in previous tables are used here.

GeoCast integrates the physical location into routing and addressing in the network design and provides effective group communication to a geographic region. The hierarchical arrangement of GeoRouters based on the nested service areas reduces the size of the routing tables. LAR inherits the bandwidth saving of on-demand routing when there is no data to send. Moreover, it reduces DSR overhead by restricting

Table 5.3   Characteristics of GPS assisted routing.

	GeoCast	LAR	DREAM	GPSR
Support location propagation	Yes	Yes	Yes	No
Data forwarding by location	Yes	No	Yes	Yes
Routing philosophy	Proactive	On-demand	Proactive	Proactive (beacons only)
Sensitive to mobility	No	Yes	No	No
Routing metric	Shortest path	Shortest path	Shortest path	Closest distance
Loop-free	Yes	Yes	Yes	No
Worst case exists	No	Yes (full flooding)	No	Yes (loops and longer paths)
Multiple receivers	Yes	No	No	No
Storage complexity	$O(N)$	$O(N)$	$O(N)$	$O(M)$
Comm. complexity	$O(N)$	$O(e)$	$O(N)$	$O(M)$

the propagation of route request packets. However, when no path is available within the limited request zone or when location information is obsolete, LAR reverts to DSR's full area flooding. Geographic information is used only in flood reduction during route discovery. DREAM adopts a pure proactive approach for location updates at each node. It makes data forwarding decisions based on the geographic information carried by the data packet. Partial flooding of the data packet toward the direction of the destination results in multipath forwarding of copies of the original packets to the destination. This multiple delivery increases the probability of reception and protects DREAM from mobility. Both LAR and DREAM involve network-wise flooding to obtain location information. Thus, the control overhead increases when the network grows.

GPSR decouples geographic forwarding from location services. The routing overhead is limited to only periodic beacon massages and a small table for neighbor locations (compared to GeoCast and DREAM, where tables contain all the nodes in the network). Thus, GPSR achieves its scalability by being insensitive to the number of nodes in the network. However, additional overhead for location services (including location registration and location database lookup) must be considered when GPSR is used. Overhead is usually restricted because only destinations need to register to the location database and only sources need to query the database. Also, lookup is performed only once at the time communication starts. Ongoing connections will exchange location updates through the data packet headers. A scalable location lookup scheme can be found in Li et al.[46]

5.5   The MINUTEMAN Project

In the previous section several advanced, scalable routing protocols were presented. In the sequel, we will apply such protocols to the **automated, digitized battlefield** scenario. We already introduced this scenario in Section 5.2.1. In this section we further elaborate on it by providing a case study of automated battlefield networking in the context of the MINUTEMAN project at UCLA, Henry Samueli School of Engineering and Applied Sciences. This project was funded by the Office of Naval Research from 2000 to 2005 and was an ideal testing ground for scalable routing schemes. The main goal of the MINUTE-MAN project was to develop the concept and initial prototype of an agile, dynamic Internet in the Sky architecture that could support the demanding communications requirements of the agents and could deliver the "forward power" of the unmanned missions. Here, we briefly describe the challenges that such an Internet in the Sky design poses in the face of the unique requirements of the unmanned missions. We also outline the innovative solutions developed to meet such challenges. Incidentally, many of the results reported here which relate to autonomous networking, scalable routing, UAV/UGV backbones, resilience to attacks, and real-time traffic support are directly applicable also to **Homeland Defense** scenarios and **Urban Warfare** situations.

The first challenge is to handle **agent mobility**, which may vary from the roving speed of the UAGs all the way to the hundreds of miles per hour speed of airborne assets (UAVs) during an attack mission. The traditional Mobile IP approach will not scale to a large number of mobile agents, high speeds, and pervasive mobility: the registration of the mobile with Foreign Agents introduces excessive overhead, and the rerouting via Home Agent and Foreign Agent becomes impractical. The approach is to embed mobility support at OSI layer 2, using ad hoc networking and ad hoc routing, below IP (we still retain, however, the Mobile IP paradigm for communications with the wired Internet). Moreover, we exploit the fact that agents typically move in and achieve scalability by keeping track of group rather than individual movements. The scalable approach to group mobility management and routing has been implemented in the LANMAR architecture.[18] For the MBN, the LANMAR architecture has been extended to large node populations and large geographical distances using multilayering (i.e., backbone) concepts. LANMAR results for representative scenarios will be shown. Moreover, LANMAR will be shown to maintain robust, resilient, rapidly restored, nearly optimal (in terms of path length) connectivity in the face of agent mobility.

The robust, all-time connectivity provided by LANMAR is critical, but it is not sufficient to carry out successful missions. The UAVs gathering intelligence at the forefront must be able to transmit multimedia (e.g., compressed video) streams with bandwidth guarantees across the backbone network to other clusters of mobile agents preparing the attack or to the commander on the ship. Thus, a second important challenge for the airborne Internet is to support **QoS** in terms of bandwidth, response time delay, and delay variation. This project has taken a multilayer approach to QoS that includes backbone beam forming at the radio layer, MAC layer scheduling, network layer QoS routing, Call Acceptance Control, and backbone path pinning by means of label switched paths and Multi Protocol Label Switching (MPLS). MPLS is used only on the relatively stable backbone. It provides the flexibility to forward individual and/or aggregated flows on QoS compliant multiple paths selected by the QoS routing algorithm, overcoming the limitations of traditional shortest path routing. These concepts are implemented in the "MBN" architecture that builds upon the LANMAR connectivity management and provides QoS where needed. QoS support requires the allocation and "alignment" of several network resources (e.g., backbone UAVs in strategic positions). If the UAVs are destroyed or reassigned to a more critical mission, QoS will be gracefully degraded.

A third critical requirement is to dynamically **adjust to environment changes** that are due either to natural causes (e.g., radio propagation irregularities, fading, mobility, obstacles, battery power depletion, etc.) or to adversary actions (e.g., UAV destruction, radio jamming, etc.). In view of such abrupt and often unpredictable changes, network protocols and applications must react in concert and must adaptively readjust to the new situation. We will discuss in the sequel various adaptive protocol features (both intra- and interlayer) that were designed specifically to address these changes. Moreover, the total unpredictability of these changes makes it impossible to provide "guaranteed" QoS, as it is generally done in commercial networks. Instead, the concept of guaranteed QoS is replaced by that of "adaptively renegotiable" QoS. Particular attention is given in this project to the adjustment of compressed video (say, MPEG 4) parameters in order to make the best use of the existing network resources.

Dynamic adaptation is also required in the assembly of resources to launch a mission and to track its progress. In this respect, the unique feature of the unmanned agent system is that some agents can support multiple functions. For example, an UAV can be used for communications, as a node of the MBN as well as for intelligence, to gather video and images as part of a scouting mission. Thus, planning a mission requires the allocation of limited resources and possibly the "reallocation" of resources from background missions to top priority missions. Monitoring and dynamic reallocation of resources based on time changing priorities, along with QoS renegotiation, are performed by a distributed, systems wide Adaptive Resource Monitoring and Management System. This system permits the various battlefield resources (from communications to CPU power, memory, and databases) to dynamically reallocate across multiple simultaneous missions in the most efficient manner.

Advanced applications such as Automatic Target Recognition (ATR) require the gathering of video and sensor information from vast areas in the battlefield in order to determine the presence and type of targets. This information must be received with extremely tight accuracy and time constraints in order, say, to execute a successful strike mission. Brute force scanning of the entire area may not be feasible; it may require too much time given the available UAV and sensor assets. In order to accomplish the goal within the required constraints, a

distributed information database provides global information about assets in the battlefield, as well as video and images captured during routine surveillance. The information database can effectively "guide" UAV teams in the search of critical areas. It also can supplement the UAV and sensor image data with stored information, thus reducing the time to target detection and recognition. The maintenance of a timely and accurate information database in the battlefield poses several new challenges, including a distributed, fault tolerant implementation; the ability to answer queries with a variable degree of accuracy depending on time constraints; and a careful tradeoff between the background refresh rate (and thus accuracy) and the use of limited communications and sensor resources.

The dynamic adaptation to unpredictable, hostile environments requires the support of advanced, **programmable radios** and of adaptive modulation and channel encoding schemes. The goal here is to achieve the best use of the available spectrum while providing the radio range, beam directivity, and channel quality required by the upper protocol layers. An important contribution of this project is the development of "modular" radios that utilize advanced MIMO and OFDM techniques and that can be dynamically reconfigured to fit the needs of an extremely broad range of platforms, from low power stationary sensors to fast flying UAVs with video capture and transmission.

Finally, the demonstration of a highly adaptive suite of protocols is itself a challenge. It does not suffice to demonstrate each component in isolation; the key is the successful interoperation of the components and the cooperative, interlayer adaptation to unpredictable changes in the environment. To this end, the MINUTEMAN project leverages a **"hybrid" simulator capability** that allows a widely ranging set of configuration parameters (number of nodes, speeds, etc.) to interface "real" applications to simulated innercore network protocols. The hybrid simulation testbed is an essential "expander" of the hardware testbed, which is by practical necessity limited in number, speed, and geographic scope.

In summary, the adaptive, unmanned agent Internet in the Sky project requires an unprecedented degree of adaptivity in the design of the various protocol layers, from radio to applications, and the development of new adaptive middleware and new hybrid simulation techniques for testbed deployment and evaluation. As the trend in modern communications systems, both military and civilian, is to become increasingly more complex, autonomous, and "adaptive," the

MINUTEMAN innovative solutions can be effectively transferred to several related application domains.

5.6   Scalable Routing in MINUTEMAN

Usually, a MANET is assumed to be flat and homogeneous. However, a flat ad hoc network has poor scalability.[1,2,11] In Gupta and Kumar,[1] theoretical analysis implies that even under the optimal circumstances, the throughput for each node declines rapidly toward zero, while the number of nodes is increased. This is proved in an experimental study of scaling laws in ad hoc networks employing IEEE 802.11 radios presented in Gupta, Gray, and Kumar.[2] The measured per node throughput declines much faster in the real testbed than in theory. Simulation results in Das, Perkins, and Royer[10] also demonstrated that while routing protocols are applied, their control overhead consumes most available bandwidth when the traffic is heavy. Besides limitation of available bandwidth, the "many hop" paths in a large-scale network are prone to break and cause many packet drops. Packet drop can be treated as a waste of bandwidth and can worsen network performance. All these issues prevent the flat ad hoc network from scaling to large scale. Thus, a new methodology is needed for building a large-scale ad hoc network. An emerging promising solution is to build a physically hierarchical ad hoc network and mobile wireless backbones.

The hierarchical ad hoc network structure employed in MINUTE-MAN makes use of mobile backbones (MBN). A general illustration of a two-level MBN is given in Figure 5.8. Among the mobile nodes, some nodes, named backbone nodes (BNs), have an additional powerful radio to establish wireless links among themselves. Thus, they form a higher

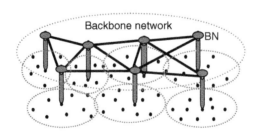

Figure 5.8   General model of a two-level MBN.

level network called a backbone network. Since the BNs are also moving and join or leave the backbone network dynamically, the backbone network is just another hoc network running in a different radio level. Multiple level MBNs can be formed recursively in the same way.

Three critical issues are involved in building such an MBN: (1) the optimal number of BNs, (2) BN deployment, and (3) routing. Assuming that the number of BNs has been determined, the second important issue is how to deploy them in the field. The main challenges in carrying out an efficient deployment are mobility and BN failures. Using a clustering scheme to elect the BNs is a natural choice, since clustering has been widely used in the past to partition nodes into small sets and to form hierarchical networks.[6,7] However, a major drawback of current clustering schemes is cluster instability in the face of mobility.[6] Unstable clusters lead to frequent clusterhead changes and thus BN changes. The backbone topology would then be too dynamic to be tracked by routing and too unpredictable to be relied upon for QoS support. In the sequel, we present a novel, fully distributed clustering scheme that achieves good stability.

Routing also critically affects the hierarchical network performance. Simply stated, routing must utilize the wireless backbone links efficiently. The main challenge that sets wireless networks apart from the wired Internet is mobility: in an Internet-like routing scheme, address prefixes would need to be continuously changed as nodes move! The overhead associated with address management would easily offset the routing control traffic and routing table size reductions offered by the hierarchical structure. LANMAR has proven to be a very effective scheme in large networks with group mobility.[17,18] In the MINUTEMAN project, LANMAR is extended to the MBN architecture. The extended version retains the simplicity of the traditional flat scheme. Yet, it preserves all the typical backbone strategy benefits, namely, short paths to remote nodes, low end-to-end delay, high quality links, augmented network capacity, and enhanced QoS support. Moreover, LANMAR exploits the hierarchical structure by reducing control overhead and propagating routing information more promptly.

5.7 Backbone Node Deployment and Clustering

One way to deploy the backbone network is to precompute first the optimal number of BNs required by the given initial node layout.

Then, one distributes the BNs uniformly in the field at initialization. However, this two-step procedure has two problems. First, the BNs move; thus after a while, some BNs may collide or anyway interfere with each other, while some areas may be uncovered. Second, BNs may fail or even be destroyed. New BNs must be deployed to replace the failed ones. A static, a priori allocation and deployment cannot efficiently fulfill both requirements.

In MINUTEMAN the solution is to combine allocation (of number of BNs) and deployment by initially assigning *redundant* backbone capable nodes and letting the *election* procedure choose the active backbone set dynamically to meet the changing requirements. A node is backbone capable if it has the physical radio capacity to communicate with other BNs and join the backbone network. If the backbone capable nodes are redundant, i.e., are in more ample supply than strictly needed, only a subset of them joins the backbone at any given time. The remaining candidates are kept as spare nodes. When one BN is destroyed or moves out of a certain area, a new BN will be dynamically selected from the backbone capable set to replace it. If two BNs move too close to each other, one of them will give up its backbone role.

The procedure to select BNs from capable nodes is totally distributed and is called *backbone election*. It leads to the proper number of BNs that uniformly cover the entire area. Clustering has been traditionally used to select a subset of nodes.[6,7] Here, we use clustering to create a physical hierarchy. In the sequel, we briefly review some options and then introduce our solution.

5.7.1   Random Competition-Based Clustering

Many clustering schemes have been proposed in the literature.[3–6] Among them, the Lowest ID (LID) and Highest Degree (HD) algorithms are widely used due to their simplicity. The detail of the two algorithms can be found in Lin and Gerla[3] and Gerla and Tsai.[5] Previous research in clustering mainly focuses on how to form clusters with a good geographic distribution, such as minimum cluster overlap, etc. However, stability is also an important criterion, especially when clustering is used to support routing. In particular, in the MINUTEMAN hierarchical structure, stability of BNs is a must. Previous clustering schemes cannot meet such a requirement.

Targeting both stability and simplicity, we have designed a new scheme called Random Competition Clustering (RCC). The main idea is that any candidate node, which currently does not belong to any cluster, can initiate a cluster formation by broadcasting a packet to claim itself as a clusterhead. The first node, which broadcasts such a packet, will be elected as the cluster-head by its neighbors. All the immediate neighbors, after hearing this broadcast, give up their right to be a clusterhead and become members of the cluster. Clusterheads have to periodically broadcast a clusterhead claim packet to maintain their status. Since there is a delay from when one node broadcasts its clusterhead claim packet to when this packet is heard by its neighbors, several neighbor nodes may broadcast during this period. To reduce such concurrent broadcasts, we introduce a random timer. Each node defers a random time before its clusterhead claim. If it hears a clusterhead claim during this random time, it then gives up its broadcast. The idea of "first claim node wins" (independently of ID number or connectivity degree) was first proposed in the Passive Clustering scheme in Gerla, Kwon, and Pei.[8] The First Claim Wins scheme favors the Clusterhead, which can be challenged only by preexisting Clusterheads. In Passive Clustering, clusters are formed on demand, when user traffic is present. In the absence of traffic, the clusters are dissolved. Our scheme is active clustering (as the election is carried out continuously in the background), but we nevertheless use the same concepts of "first declaration" and explicit random timer. Of course, the random timer cannot completely solve the concurrent broadcast problem. When concurrent broadcast happens, we use the node ID to solve the conflict. The node with the lower ID will become the clusterhead.

The RCC scheme is more stable than traditional clustering schemes such as LID and HD. In the LID scheme, when the clusterhead hears a node with a lower ID, it will immediately give up its clusterhead role. Similarly, in the HD scheme, when a node acquires more neighbors, the cluster will also be reconfigured. Due to node mobility, such events happen very frequently. In RCC, one node only gives up its clusterhead position when another clusterhead moves near to it. Since clusterheads are usually at least two hops away, clusters formed by RCC are much more stable.

The low control overhead of RCC is clear. In the LID and HD clustering schemes, each node has to know the complete information of neighbor nodes. In our scheme, only the clusterheads need to broadcast a small control packet periodically. All other nodes just keep silent.

5.7.2 Multihop Clustering

Usually, the clustering schemes are one-hop based, that is, the clusterhead can reach all members in one hop. This is not suitable for BN election. One wants to control (and, in fact, optimize) the number of elected BNs. To achieve this, we extend the basic clustering scheme so that it creates K-hop clusters. Here, K hop means that a clusterhead can reach any one of its members in at most K hops. By adjusting the parameter K, we can approximately control the number of clusterheads. Bigger K means fewer clusterheads and thus fewer BNs.

In K-hop clustering, each node forwards the "claim" of its clusterhead. From the claims received from the neighbors, a mobile node will select the nearest clusterhead within K-hop scope. If there is no clusterhead within a K-hop scope, a node claims itself as a clusterhead after deferring for a random time. In K-hop clustering, the probability of competing clusterhead claims is relatively high due to the longer time for propagating clusterhead claim packets K hops away. The random time delay helps break ties.

5.8 Scalable Routing: LANMAR

After the BNs are elected, powerful backbone radios are used to connect BNs and form a backbone network. Now, the critical issue is routing. The backbone links among BNs provide "short cuts" and high bandwidth. Routing must be able to exploit backbone links for remote destinations. Moreover, since BNs may fail or even be destroyed, routing must be reliable and tolerant of such failures. The solution is based on the LANMAR scheme introduced in Section 5.4.2.[17,18] Here, we extend LANMAR to include also the MBN, yet preserve its scalability and fault tolerant properties.

5.8.1 LANMAR in the MBN

In the original LANMAR scheme, a packet is routed toward the corresponding remote landmark along the (typically long) multihop path advertised by the DV algorithm. The new scheme includes backbone links. In this scenario, the min hop path will generally include some of the backbone links. Thus, the basic LANMAR automatically and "opportunistically" makes use of the backbone without requiring any specific modifications. In practice, the packet is routed to the nearest BN. This local BN then forwards the packet to a remote BN near the destination landmark via the backbone. Finally, the remote BN delivers the packet to the remote landmark or directly to the destination if it is within its Fisheye scope. This greatly reduces the number of hops. This procedure is illustrated in Figure 5.9. We can see that by utilizing the backbone links, the 6 hop path is reduced to be 3 hops, a great improvement! The landmarks are mapped to BNs (multiple mappings are possible for fault tolerance). The route within the MBN is computed by the MBN unique routing algorithm and may, in fact, satisfy given QoS constraints.

A possible implementation of the MBN/LANMAR scheme is as follows. First, all mobile nodes, including BNs, run the original LANMAR routing on the "local" links via short-range radios. This is the foundation for falling back to flat multihop routing if the MBN fails. Second, a BN periodically broadcasts its landmark map (i.e., the landmark distance vector) to neighbor BNs via the backbone links. The neighbor BNs treat this packet as a normal landmark update packet. Since this higher level path is usually shorter, it replaces the long multihop paths. From landmark updates the ordinary nodes thus learn the best path to the remote landmarks, including the paths that utilize the backbone links. Each BN needs to record the radio interface to the next hop on each advertised path in order to route packets through the correct radios later. As discussed earlier, the routing within the MBN need not be "shortest path"; it may, in fact, be QoS routing.

Important features of the proposed routing scheme are reliability and fault tolerance. The ordinary nodes are prevented from knowing the backbone links explicitly. The backbone links are automatically

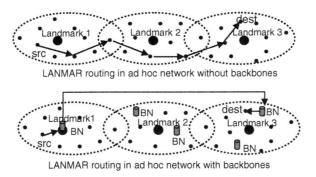

LANMAR routing in ad hoc network without backbones

LANMAR routing in ad hoc network with backbones

Figure 5.9   LANMAR routing in MBN.

learned via routing broadcasts from BNs. Now, if a BN is destroyed, the shorter paths advertised by this BN will soon expire. Then, new landmark information broadcast from other nodes will replace the expired information. Thus, in the worst case, routing in this cluster goes back to original landmark routing while other groups with BN support can still benefit from the backbone "short cuts." When all backbone capable nodes are disabled, the whole network again becomes a flat ad hoc network running the original LANMAR routing, which can still provide connectivity, albeit with lower performance (longer paths, no QoS support).

In the past, we have described a very simple DSDV routing scheme for the MBN, with omnidirectional antennas, neighbor discovery, and DV routing support to landmarks. This scheme is sufficient to provide short cut benefits across the backbone. More elaborate and efficient MBN configurations (e.g., point-to-point links) and routing schemes (e.g., LS) are currently being investigated in the MINUTEMAN project. With MBN LS routing, each BN can advertise the landmarks within its reach. A BN can then select the most cost-effective route within the backbone to the intended destination. This makes QoS support possible.

5.8.2   Mobile IP Routing from the Internet

LANMAR is a MANET routing protocol. As such, it only supports routes inside the ad hoc network. In order to route packets to/from the Global Internet, other mechanisms are required. In particular, each node is assigned an IP address (IPv4 or IPv6). This IP address is used by corresponding nodes in the Global Internet to reach our mobile node using Mobile IP via the Home Agent and the Foreign Agent (FA) architecture. The FA is effectively a Name Server that provides the mapping from IP address to LANMAR address. Mobile nodes refresh the Name Server database periodically. They refresh immediately if they join a new group. In IPv6 the local address field becomes the MANET address, namely, the LANMAR address. This helps remove the inefficiencies of Mobile IPv4 (tunneling and triangular routing).

5.9   Simulation Experiments

In this section, we evaluate the efficiency of the clustering and routing solutions so far proposed. We use GlomoSim/Qualnet,[16] a packet level network simulation platform for ad hoc networks based on the parallel language PARSEC. We begin with the RCC algorithm. We compare the stability of the RCC algorithm with the LID and HD algorithms. Since we are targeting large-scale networks, we deploy 1000 mobile nodes in a "terrain" of size $3200 \times 3200\,$m. Each mobile node has an IEEE 802.11 wireless radio with a transmission range of 175 m. The Distributed Control Function (DCF) mode of IEEE 802.11 is used, and the channel bandwidth is set to 2 Mbps. The node mobility model is random waypoint mobility.[14] In the simulation experiments, the pause time is kept constant and equal to 30 seconds while speed is varied to observe the stability of the clusters. The simulation time is 6 minutes for all runs.

The stability of clusters includes two parts: the stability of the clusterhead and the stability of the cluster members. These are measured by two different metrics: average lifetime of a clusterhead and average membership time of a cluster member. In the MBN, the average lifetime of a clusterhead is exactly the average lifetime of a BN. In this simulation experiment, we only implement the basic clustering scheme without considering the "gateway" node selection as in Lin and Gerla[3] and Gerla and Tsai.[4]

5.9.1   Cluster Stability

Usually, clustering is performed to form one-hop clusters. Thus, here we compare the stability of one-hop clusters. Again, one hop means that the clusterhead can reach all its members in one hop. The simulation results are given in Figures 5.10 and 5.11.

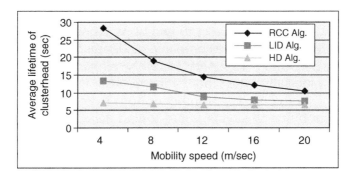

Figure 5.10   Average lifetime of clusterhead.

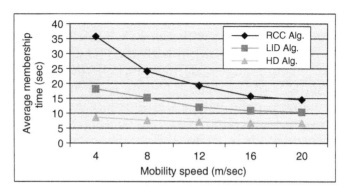

Figure 5.11 Average membership time.

From Figures 5.10 and 5.11, the RCC algorithm is more stable than the LID and HD algorithms in both low mobility and high mobility scenarios.

5.9.2 Routing Algorithm Performance

In this section, we compare the LANMAR extension in MBN with the original LANMAR routing and AODV,[13] a popular reactive routing protocol in the flat ad hoc network. The basic environment is kept the same as in the clustering experiment, i.e., 1000 mobile nodes. Each ordinary node has a small 802.11 wireless radio with a power range of 175 m and a channel bandwidth of 2 Mbps. The BNs have two 802.11 radios: one small radio the same as the ordinary nodes and one powerful radio with a power range of 800 m and a channel bandwidth of 5 Mbps. The mobility model is "group mobility" as presented in Hong et al.[19] Thirty Continuous Bit Rate (CBR) source-destination pairs on top of UDP (User Datagram Protocol) are used to generate the traffics. The scope of backbone election is set to two hops. Node speed is increased from 0 to 10 m/sec. Results are compared in Figures 5.12 and 5.13.

In Figures 5.12 and 5.13, LANMAR/MBN outperforms flat LANMAR and AODV, especially when nodes move. This is because it utilizes backbone links to reduce the number of hops from sources to destinations. With mobility, the average end-to-end delay of AODV is greatly increased. This is due to the reactive feature of AODV. For increasing speed, links break and the path expires more frequently. AODV then must delay packets as it searches for a new path from source to destinations. In contrast, LANMAR and LANMAR/MBN are proactive.

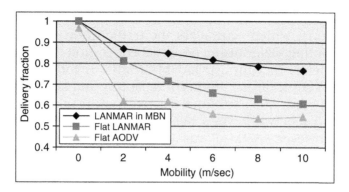

Figure 5.12 Comparison of delivery fraction in mobility.

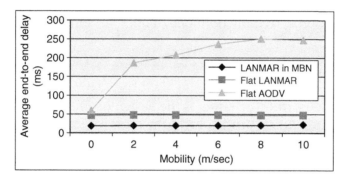

Figure 5.13 Comparison of end-to-end delay in mobility.

Thus, their delay is not affected significantly by speed. LANMAR in MBN further reduces the delay using backbone links.

5.10   Related Work

The MINUTEMAN project has introduced a new concept in scalable ad hoc network design, namely, the use of an MBN. Embedding the backbone network in the original large-scale ad hoc network requires two new provisions. One is the "clustering" scheme to select the number and location of the backbone nodes. The other is the routing in the backbone. Prior work in both areas was reported in the literature. Below we contrast the MINUTEMAN contributions to prior work.

In Gu et al.,[12] routing in the UAV-based hierarchical structure is investigated. Clustering is also used to select BNs. However, only one-hop clustering is used (as opposed to K-hop clustering in

MINUTEMAN). The routing scheme is fully folded onto the hierarchical structure, which centralizes the traffic from each cluster into the corresponding BN, causing potential congestion and single-point-failure problems. In contrast, LANMAR/MBN is fully decentralized, with obvious advantages in terms of reliability and fault tolerance.

In Ko and Vaidya,[9] a traditional reactive routing scheme is extended to a hierarchical network. The scheme does provide reliability and fault tolerance. However, LANMAR was shown to possess several advantages over flat reactive routing. These advantages still persist when LANMAR/MBN is implemented in a hierarchical structure. For example, the reactive hierarchical scheme inherits the long delay of new path discovery. Path discovery delay and end-to-end delay (for data packets) will keep increasing, especially in high mobility. In contrast, the LANMAR/MBN is proactive and bounded. It does not suffer such problems.

5.11   Conclusion

The main theme of this chapter is the design of ad hoc networks, with a focus on the routing protocol and the need for scalable, robust solutions. In the first part of the chapter, we reviewed a broad range of ad hoc routing protocols and investigated their scalability and robustness. We provided detailed descriptions and discussed the differences, highlighting particular features that impact performance. No protocol emerges as the winner for all scenarios. Rather, each protocol offers different, competitive, and complementary advantages that make it suitable for different applications and implementations. Indeed, this is very appropriate given the current proliferation of ad hoc network applications, from battlefield to homeland security, search and rescue, personal/group communications, and sensor networks, all with very diversified requirements. In the future, researchers will face even more challenges in the attempt to find the best match between the properties of routing schemes on one hand and the network environment (radio and MAC configurations, propagation effect, interference, etc.) and user demands on the other hand. In this extremely diverse scenario there will be a need for routing protocols that are flexible and capable of adapting to various population scales, traffic types and patterns, security needs, and application domains. Some initial proposals in this direction are emerging, for instance, routing protocols that combine proactive and reactive features and that can exploit geo-coordinates.

In the second part of the chapter, we introduced the MINUTE-MAN Internet in the Sky "case study." This is a very significant study as it is representative of the three major applications of ad hoc networking today, namely, battlefield, disaster recovery, and Homeland Defense. We reviewed the critical issues involved in the development of a scaleable routing protocol and in the deployment of an MBN. The key novelty was, in fact, the presence of the MBN that must be transparently discovered and properly exploited by routing. We proposed a new stable clustering scheme to deploy the BNs. We also proposed a LANMAR/MBN routing extension that operates efficiently and transparently with the backbone network. Backbone links are automatically selected by the routing scheme if they can reduce hop distance to remote destinations. Fault tolerance and system reliability are also targeted in the design, with very satisfactory results. In essence, the proposed scheme combines the benefits of flat LANMAR routing and physical network hierarchy. Simulation results using Parsec/GloMoSim platform show that the proposed scheme can establish and operate an MBN effectively and efficiently. It can improve the network performance significantly, and it is robust to failures.

References

1. P. Gupta and P. R. Kumar, "The Capacity of Wireless Networks," *IEEE Transactions on Information Theory*, vol. IT-46, no. 2, pp. 388–404, March 200.
2. P. Gupta, R. Gray, and P. R. Kumar, "An Experimental Scaling Law for Ad Hoc Networks," Univ. of Illinois at Urbana-Champaign, Technical Report, May 2001.
3. C. R. Lin and M. Gerla, "Adaptive Clustering for Mobile Networks," *IEEE Journal on Selected Areas in Communications*, vol. 15, no. 7, pp. 1265–1275, September 1997.
4. M. Gerla and J. T. Tsai, "Multicluster, Mobile, Multimedia Radio Network," *ACM-Baltzer Journal of Wireless Networks*, vol. 1, no. 3, pp. 255–265, 1995.
5. P. Krishna, N. H. Vaidya, M. Chatterjee, and D. K. Pradhan, "A Cluster-Based Approach for Routing in Dynamic Networks," in Proceedings of ACM SIGCOMM Computer Communication Review '97, pp. 372–378, Cannes, France, 1997.
6. S. Banerjee and S. Khuller, "A Clustering Scheme for Hierarchical Control in Multi-Hop Wireless Networks," in Proceedings of IEEE INFOCOM '01, Anchorage, Alaska, April 2001.

7. P. Sinha, R. Sivakumar, and V. Bharghavan, "Enhancing Ad Hoc Routing with Dynamic Virtual Infrastructures," in Proceedings of IEEE INFOCOM '01, Anchorage, Alaska, April 2001.

8. M. Gerla, T.J. Kwon, and G. Pei, "On Demand Routing in Large Ad Hoc Wireless Networks with Passive Clustering," in Proceedings of IEEE WCNC '00, Chicago, IL, September 2000.

9. Y. Ko and N. H. Vaidya, "A Routing Protocol for Physically Hierarchical Ad Hoc Networks," Texas A&M University, Technical Report 97–010, September 1997.

10. S. R. Das, C. E. Perkins, and E. M. Royer, "Performance Comparison of Two On-Demand Routing Protocols," in Proceedings of IEEE INFOCOM '00, Tel Aviv, Israel, March 2000.

11. M. S. Corson, "Flat Scalability — Fact or Fiction?" in Proceedings of ARO/DARPA Workshop on Mobile Ad Hoc Networking '97, University of Maryland, College Park, March 1997.

12. D. L. Gu, G. Pei, M. Gerla, and X. Hong, "Integrated Hierarchical Routing for Heterogeneous Multi-Hop Networks," in Proceedings of IEEE MILCOM '00, Los Angeles, CA, October 2000.

13. C. E. Perkins and E. M. Royer, "Ad-Hoc On-Demand Distance Vector Routing," in Proceedings of IEEE WMCSA '99, New Orleans, LA, February 1999.

14. D.B. Johnson and D A. Maltz, "Dynamic Source Routing in Ad Hoc Wireless Networks," in *Mobile Computing* (ed. T. Imielinski and H. Korth), Dordrecht/Norwell, MA: Kluwer Academic, Ch. 5, pp. 153–181, 1996.

15. C. Perkins and P. Bhagwat, "Highly Dynamic Destination-Sequenced Distance-Vector Routing (DSDV) for Mobile Computers," in Proceedings of the ACM SIGCOMM '94, London, UK, October 1994.

16. M. Takai, L. Bajaj, R. Ahuja, R. Bagrodia, and M. Gerla, "GloMoSim: A Scalable Network Simulation Environment," UCLA, Computer Science Department, Technical Report 990027, 1999.

17. M. Gerla, X. Hong, and G. Pei, "Landmark Routing for Large Ad Hoc Wireless Networks," in Proceedings of IEEE GLOBECOM '00, San Francisco, CA, November 2000.

18. G. Pei, M. Gerla, and X. Hong, "LANMAR: Landmark Routing for Large Scale Wireless Ad Hoc Networks with Group Mobility," in Proceedings of IEEE/ACM MobiHOC '00, Boston, MA, August 2000.

19. X. Hong, M. Gerla, G. Pei, and C. C. Chiang, "A Group Mobility Model for Ad Hoc Wireless Networks," in Proceedings of ACM/IEEE MSWiM '99, Seattle, WA, August 1999.

20. G. Pei, M. Gerla, and T. W. Chen, "Fisheye State Routing in Mobile Ad Hoc Networks," in Proceedings of the 2000 ICDCS Workshops, Taipei, Taiwan, April 2000.

21. S. Keshav, *An Engineering Approach to Computer Networking: ATM Networks, the Internet, and the Telephone Network*. Reading, MA: Addison–Wesley

Professional Computing Series, Addison–Wesley Longman Publishing, Ch. 11, 1997.

22. K. Xu, X. Hong, and M. Gerla, "An Ad Hoc Network with Mobile Backbones," in Proceedings of IEEE ICC '02, New York, April 2002.

23. S. R. Das, R. Castaneda, and J. Yan, "Simulation Based Performance Evaluation of Mobile, Ad Hoc Network Routing Protocols," *ACM/Baltzer Mobile Networks and Applications (MONET) Journal*, pp. 179–189, July 2000.

24. E. M. Royer and C.-K. Toh, "A Review of Current Routing Protocols for Ad-Hoc Mobile Wireless Networks," *IEEE Personal Communications*, pp. 46–55, April 1999.

25. C. Santivanez, R. Ramanathan, and I. Stavrakakis, "Making Link-State Routing Scale for Ad Hoc Networks," in Proceedings of the 2nd ACM International Symposium on Mobile Ad Hoc Networking & Computing, Long Beach, CA, October 2001.

26. Z. J. Haas and M. R. Pearlman, "The Performance of Query Control Schemes for the Zone Routing Protocol," *IEEE/ACM Transactions on Networking (TON)*, vol. 9, no. 4, pp. 427–38, August 2001.

27. J. Broch, et al., "A Performance Comparison of Multi-Hop Wireless Ad Hoc Network Routing Protocols," in Proceedings of ACM/IEEE MobiCom '98, pp. 85–97, Dallas, TX, October 1998.

28. S.-J. Lee, C.-K. Toh, and M. Gerla, "Performance Evaluation of Table-Driven and On-Demand Ad Hoc Routing Protocols," in Proceedings of IEEE PIMRC '99, pp. 297–301, Osaka, Japan, September 1999.

29. C. E. Perkins, *Ad Hoc Networking*. New York: Addison–Wesley, 2001.

30. X. Hong, et al., "Scalable Ad Hoc Routing in Large, Dense Wireless Networks Using Clustering and Landmarks," in Proceedings of IEEE ICC '02, New York, April 2002.

31. A. Iwata, et al., "Scalable Routing Strategies for Ad-Hoc Wireless Networks," *IEEE Journal of Selected Areas on Communications*, vol. 17, pp. 1369–1379, August 1999.

32. P. Jacquet, et al., "Optimized Link State Routing Protocol," draft-ietf-manetolsr-05.txt, Internet Draft, IETF MANET Working Group, November 2000.

33. A. Qayyum, L. Viennot, and A. Laouiti, "Multipoint Relaying: An Efficient Technique for Flooding in Mobile Wireless Networks," INRIA, Technical Report RR-3898, Rapport de Recherche, 2000.

34. B. Bellur and R. G. Ogier, "A Reliable, Efficient Topology Broadcast Protocol for Dynamic Networks," in Proceedings of IEEE INFOCOM '99, New York, March 1999.

35. R. G. Ogier, et al., "Topology Broadcast Based on Reverse-Path Forwarding (TBRPF)," draft-ietf-manet-tbrpf-05.txt, Internet-Draft, MANET Working Group, March 2002.

36. C.-K. Toh, "Associativity-Based Routing For Ad Hoc Mobile Networks," Wireless Personal Communications Journal, Special Issue on Mobile

Networking and Computing Systems, Kluwer Academic Publishers, vol. 4, no. 2, pp. 103–139, March 1997.

37. M. S. Corson and A. Ephremides, "A Distributed Routing Algorithm for Mobile Wireless Networks," *ACM/Baltzer Wireless Networks*, vol. 1, no. 1, pp. 61–81, February 1995.

38. V. D. Park and M. S. Corson, "A Highly Adaptive Distributed Routing Algorithm for Mobile Wireless Networks," in Proceedings of IEEE INFOCOM '97, pp. 1405–1413, Kobe, Japan, April 1997.

39. C.-C. Chiang and M. Gerla, "Routing and Multicast in Multihop, Mobile Wireless Networks," in Proceedings of IEEE ICUPC '97, pp. 546–552, San Diego, CA, October 1997,

40. G. Pei, et al., "A Wireless Hierarchical Routing Protocol with Group Mobility," in Proceedings of IEEE WCNC '99, New Orleans, LA, September 1999.

41. G. Pei and M. Gerla, "Mobility Management in Hierarchical Multi-Hop Mobile Wireless Networks," in Proceedings of IEEE ICCCN '99, Boston, MA, October 1999.

42. J. C. Navas and T. Imielinski, "Geographic Addressing and Routing," in Proceedings of the 3rd Annual ACM/IEEE International Conference on Mobile Computing and Networking, Budapest, Hungary, September 26–30, 1997.

43. Y.-B. Ko and N. H. Vaidya, "Location-Aided Routing (LAR) in Mobile Ad Hoc Networks," *Mobile Computing and Networking*, pp. 66–75, 1998.

44. S. Basagni, et al., "A Distance Routing Effect Algorithm for Mobility (DREAM)," in Proceedings of ACM/IEEE MobiCom '98, pp. 76–84, Dallas, TX, 1998.

45. B. Karp and H. T. Kung, "GPSR: Greedy Perimeter Stateless Routing for Wireless Networks," in Proceedings of ACM/IEEE MobiCom '00, Boston, MA, August 2000.

46. J. Li, et al., "A Scalable Location Service for Geographic Ad Hoc Routing," in Proceedings of ACM/IEEE MobiCom '00, Boston, MA, August 2000.

47. M. Gerla, K. Xu, and A. Moshfegh, "Minuteman: Forward Projection of Unmanned Agents Using the Airborne Internet," in Proceedings of IEEE Aerospace Conference '02, Big Sky, MT, March 2002.

6

Sensor Networks

Martin Vetterli and Răzvan Cristescu

6.1 Introduction

6.1.1 Motivation

Consider a typical sensor network scenario,[30] where sensors measure a data field (e.g., temperature) and the results of their measurements have to be transported across the network to a certain designated node called the *sink* (see Figure 6.1). This is referred to as *data gathering*, and it is a relevant problem in various sensor network settings, where information from the network, in its coded form, is needed at a central base station node for storage, monitoring, or control purposes.

We address this problem from a joint source–channel coding perspective, corresponding to the network layer. Given the statistical structure of the correlated data field, we study the interactions between the representation of the measured information and the network over which this information needs to be transmitted. More specifically, the information is represented by means of the rate allocation employed for data coding at measuring nodes and the actual placement in the field of those nodes. The transmission network is represented by its graph structure, formed by the nodes inter-connections, on which the information is transmitted to the central processor that needs to reconstruct the data field. For quantifying these interactions, we consider a set of relevant metrics: energy used for the transmission and mean-square-error (MSE) distortion of the reconstruction. On one

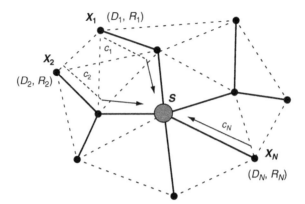

Figure 6.1 In this example, data from nodes $X_1, X_2, \ldots, X_N$ need to arrive at sink S. A rate supply R_i is allocated to each node X_i, and, in the case of lossy coding, the distortion at that node is D_i. Thick solid lines show a chosen tree transmission structure. Thin dashed lines show the other possible links. The path from node i to the sink is shown with arrows, and its weight is c_i.

hand, the *source* to be coded is the correlated data field, and the task of the coder is twofold: first, to properly sample the correlated data field so that the result is a representation with a good accuracy; and second, to allocate coding rates at the nodes such that data can be reconstructed at the sink. Thus, in our setting, the *source coder* aims at a proper representation of information at the network nodes, as a function of the characteristics of the measured field and of the nodes' positions with respect to each other. On the other hand, the *channel* is the connectivity network formed by the nodes placed in the field. The task of the *channel coder* is to find a proper transmission structure (i.e., a subset of the edges of the connectivity graph) and/or a node placement that optimizes the metric of interest. Both the source and the channel coding tasks have to be done under various restrictions, including energy and communication constraints.[*]

6.1.2 Measured Data and Network Characteristics

There are several important issues specific to sensor networks measuring and transporting data.[30] First, the measured data have certain

---

[*] Implicitly, we are using the assumption that the network is a bit-pipe and that sources will be represented in a coded digital form. This level of abstraction makes the problem tractable, but can be suboptimal in the wireless case, as demonstrated in Gastpar and Vetterli.

redundancy characteristics. For instance, if the measured data are random variables (e.g., temperature), the values at nodes are correlated, and the data structure is given by the spatio-temporal correlation.

Second, the limited coverage and transmission capabilities of the sensor nodes induce limited connectivity and communication patterns in the network graph. Nodes usually have knowledge only about other sensors situated in a limited neighborhood, so efficient joint representation of data by groups of nodes has to be done in a decentralized manner. Also, due to the battery energy limitations, most nodes cannot send their data directly to the sink, and therefore, data has to be relayed by other nodes. This implies that efficient routing is necessary, and moreover, it has to be decentralized as well. Also, depending on the coding strategy that determines the amount of internode communication, the task of data representation at nodes may or may not separate from the task of routing that data across the network.

Third, the actual node positions influence both the accuracy of the measured data and the energy efficiency of the network. For instance, placing most nodes close to the sink will improve their lifetime since they only have to transmit data over small distances. However, this will leave areas that are far from the sink uncovered, which means high inaccuracy in the overall data measurement. On the contrary, an even distribution of nodes over the measured field will provide a good data accuracy, but at the same time, the energy consumption will be large. Moreover, finding an optimized tradeoff has to take into account both data representation and routing.

Finally, in the case of data gathering of spatio-temporal processes under delay constraints, the network density influences the total distortion of reconstruction. A network with a small number of nodes results in a high spatial distortion of approximation, but the temporal distortion is small since data reaches the sink in a small number of hops. On the contrary, a network with a large density has a low spatial distortion, but the temporal distortion is high due to the large number of hops to the sink.

6.1.3   Metrics

There are certain specific metrics of interest for this type of application, namely, energy efficiency and accuracy of the data reconstruction at the sink. In sensor networks, the energy efficiency of the network depends on both the rate allocation at nodes and the routing strategy (the paths chosen to transmit the data). The energy consumed by a node is usually proportional to the product [rate] × [path weight], where the [rate] term represents the data amount (in bits) sent by a node, and the [path weight] is an increasing function of the Euclidean distance between nodes (for instance, [Euclidean distance]$^\kappa$, where $\kappa \in [2,4]$ is the path-loss exponent).

The accuracy of data reconstruction depends on the distortion allowed at measuring nodes and on the node placement and influences the data representation as well. Namely, the desired accuracy of data representation at nodes determines the rate necessary to accommodate the corresponding distortion and thus how much energy is needed to transmit that rate. Due to spatial representation reasons, the accuracy is influenced by the node placement as well. Moreover, for spatio-temporal processes, a delay represents an additional issue that affects the accuracy of the reconstruction, since the timely arrival of data reduces the total distortion.

To summarize, there is a strong interaction between data representation and routing and the metrics relevant for sensor network scenarios. This chapter presents the interaction among these important parameters for designing practical, efficient, and accurate joint measurement (source coding) and transmission (channel coding) strategies. We show that a joint consideration of the source and channel coding can result in significant improvements in terms of the metrics of interest.

6.1.4   Related Work

Bounds on the performance of networks measuring correlated data have been derived in Marco et al.,[26] and Scaglione and Servetto.[33] Progress towards practical implementation of Slepian-Wolf coding[35] has been achieved in Aaron and Girod[1] and Pradhan.[31,32] However, none of these works takes into consideration the cost of transmitting the data over the links and the additional constraints that are imposed on the rate allocation by the joint treatment of source coding and transmission.

The problem of optimizing the transmission structure in the context of sensor networks has been considered in Heinzelman[21] and Lindsey and Raghavendra,[24] where the [energy] and the [energy] × [delay]

metric are studied, and where practical algorithms are proposed. But in these studies, the correlation present in the data is not exploited for the minimization of the metric.

Recent works that exploit correlation in the data in the context of sensor networks include Barros, Peraki, and Servetto,[2] Enachescu,[15] and Pattem, Krishnamachari, and Govindan.[28].

In Pattem, Krishnamachari, and Govindan,[28] an empirical data correlation model is used for a set of experimentally obtained data, and the authors propose cluster-based tree structures shown to have a good performance depending on the correlation level. The correlation function is derived as an approximation of the conditional entropy, and the cost function is the sum of bits transmitted by the network.

In Enachescu et al.,[15] a circular-coverage correlation model on a grid is used, where correlation is modelled as a parameter proportional to the area covered by a sensor. The authors provide randomized shortest path aggregation trees with constant-ratio approximations.

Some examples of network flow with joint coding of correlated sources under capacity constraints on the transmission links and Slepian-Wolf constraints on the rates are studied in Barros, Peraki, and Servetto,[2] where trees are shown to perform suboptimally if splittable flows are allowed.

We note that, in some scenarios, uncoded transmission is optimal.[18]

A joint treatment of data aggregation and transmission structure is considered in Goel and Estrin,[20] but the model does not take into account possible collaborations for joint coding among nodes. We consider scenarios where joint source coding among nodes exploits internode correlations by means of Slepian-Wolf coding, and we compare this approach with coding by explicit communication, where such collaborations are not allowed, and where coding is done by using only available side information.

The rate-distortion region of coding with high resolution for arbitrarily correlated sources has been found in Zamir and Berger.[37] We focus on finding the optimal rate-distortion operation point when, additionally, energy constraints are imposed.

An analysis of the impact of data irregularity on the spatio-temporal sampling is done in Ganesan et al.,[17] Our novel take on the problem of data gathering of spatio-temporal processes is that we are able to formulate the problem in terms of a unique performance measure, namely, the total distortion.

6.1.5   Main Contributions and Organization of the Chapter

The main contribution of this chapter is an unified treatment of data representation, routing, and node placement in sensor networks for the optimization of various metrics of interest. First, we consider energy efficiency, and we show that, for data gathering of spatially correlated processes, the task of data representation at nodes separates from the transmission structure optimization. By using this separation result, we are able to find the optimal transmission structure and the corresponding rate-distortion allocation. Second, we consider the distortion given by the total MSE of representation of the field at the sink for energy efficient data gathering, which includes spatial and temporal distortions. We show that for spatio-temporally processes, in general, there is an optimal network density that minimizes this distortion.

In Section 6.2 we review some of the main concepts and methods used in this chapter. In Section 6.3 we present the network, transmission, and signal models analyzed in this work. Section 6.4 studies data gathering of spatially correlated, but temporally independent and identically distributed (i.i.d.) random processes, namely, the cases of lossless and high-resolution lossy coding, and addresses the node placement problem. Section 6.5 studies data gathering of spatio-temporally correlated processes. A performance measure of interest is defined, namely, the total distortion; an analysis of this measure for the one-dimensional grid is performed; and the model is generalized for a two-dimensional grid. We conclude with Section 6.6.

6.2   Background

In this section we review some of the main concepts used in this chapter.

The entropy is a measure of uncertainty of a random variable:[6] $H(X) = -\sum_{x \in \mathcal{X}} p(x) \log p(x)$, where $\mathcal{X}$ is the discrete alphabet of X, and $p(\cdot)$ is the probability distribution of X. Further, denote by $H(X|Y)$ the conditional entropy of a random variable X given that the random variable Y is known.

Consider the problem of lossless coding of two random sources X_1 and X_2 that are correlated. Intuitively, each of the sources can code their data at a rate equal to at least their corresponding entropies, $R_1 = H(X_1)$, $R_2 = H(X_2)$, respectively. If they are able to communicate, then they can coordinate their coding and use together a total rate of $R_1 + R_2 = H(X_1, X_2)$. Slepian and Wolf[35] showed that two correlated sources can be coded with a total rate equal to the joint entropy $H(X_1, X_2)$ even if they are *not* able to communicate with each other, as long as their individual rates are at least equal to the conditional entropies, $H(X_1|X_2)$ and $H(X_2|X_1)$, respectively (the resulting Slepian-Wolf rate region, formed by the achievable rate pairs (R_1, R_2), is shown in Figure 6.2). This easily generalizes to the N-dimensional case.

When coding is done in a lossy manner, that is, the sources are coded under distortion constraints, then the problem of finding the rate-distortion region becomes difficult,[3] and for most scenarios it remains open. However, a recent result[37] finds the rate-distortion region for high-resolution coding. The main idea of the proof in Zamir and Berger[37] is that at high rates, quantization followed by Slepian-Wolf coding is optimal. In the case of two sources coded with high resolution, the rate-distortion region is similar to the Slepian-Wolf region, with the addition of terms related to the MSE distortions D_1, D_2:

$$R_1 \geq h(X_1|X_2) - \log 2\pi e D_1,$$
$$R_2 \geq h(X_2|X_1) - \log 2\pi e D_2,$$
$$R_1 + R_2 \geq h(X_1, X_2) - \log(2\pi e)^2 D_1 D_2,$$

where $h(X)$ is the differential entropy of a continuous amplitude random variable.[6]

Consider now the case where sources have to be sent over a transmission channel to a certain destination or receiver. One important problem in this case is *if* and *how* to adapt the characteristics of the

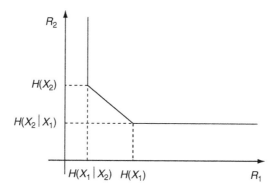

Figure 6.2 The Slepian-Wolf region for two correlated sources X_1 and X_2.

source to those of the channel. This is called joint source–channel coding, and it is a subject that has been extensively studied in the literature.[4,27] An important result of information theory states that in the point-to-point case (a single source transmitting to a single receiver) and for infinitely long block lengths, the separation of the source coder from the channel coder is optimal.[34,36] However, in the multiuser case, which includes the scenario discussed in this chapter, separation of the source and channel coding may no longer be optimal. The results in this chapter reiterate this paradigm: a joint consideration of the source coder (rate and distortion allocation at the measuring nodes) and of the channel coder (the actual transmission structure used to transport the measured information to the destination) does improve the performance of the system, in terms of both energy used for communication and accuracy of the reconstruction.

6.3   Problem Setting

6.3.1   Network Model

Consider a network of N nodes. Let $\mathbf{X} = (X_1, \ldots, X_N)$ be the vector formed by the values representing the sources measured at nodes $1, \ldots, N$. The information measured at the nodes has to be transmitted through the links of the network to the designated base station (see Figure 6.1). We assume that the interference among nodes is negligible, and there are no capacity constraints on the links (the case of omnidirectional interfering wireless channels is beyond the scope of this chapter). Such assumptions are realistic in the case of wired networks or if the antennas are unidirectional. For such scenarios, the optimal gathering structure is a tree.

In some parts of this chapter, for the sake of simplicity, we use the one-dimensional network model in Figure 6.3 rather than the two-dimensional model in Figure 6.1.

Figure 6.3   One-dimensional grid network.

6.3.2 Signal Model

6.3.2.1 Spatially Correlated Gaussian Random Field
We consider first the case where $\mathbf{X} = (X_1, \ldots, X_N)$ is a vector formed by temporally i.i.d random variables representing the sources measured at nodes $1, \ldots, N$. The samples taken at the nodes are spatially correlated and independent in time. We assume that the random variables are continuous and that there is a high-resolution quantizer in each sensor. In the lossless source coding case, a rate allocation $\{R_i\}_{i=1}^N$ (bits) has to be assigned at the nodes. In addition, if the data can be transmitted in a lossy manner, a distortion allocation $\{D_i\}_{i=1}^N$ has to be assigned as well, so that the quantized measured information samples are described with certain total distortion D and individual $\{D_i^{\max}\}_{i=1}^N$ distortion constraints.

For the sake of clarity, we use as an example a zero-mean *jointly Gaussian model* $\mathbf{X} \sim \mathcal{N}^N(0, \mathbf{K})$, with unit variances $\sigma_{ii} = 1$:

$$f(\mathbf{X}) = \frac{1}{\sqrt{2\pi} \det(\mathbf{K})^{1/2}} e^{-\left(\frac{1}{2}(\mathbf{X})^T \mathbf{K}^{-1}(\mathbf{X})\right)},$$

where $\mathbf{K}$ is the covariance matrix of $\mathbf{X}$, with elements depending on the distance between the corresponding nodes (e.g., $K_{ij} = \exp(-cd_{i,j}^\beta)$, where $d_{i,j}$ is the distance between nodes i and j,[7,9] $\beta \in \{1, 2\}$ with $\beta = 1$ corresponding to a Gauss-Markov field or $\beta = 2$ corresponding to a squared distance correlation model, and $c > 0$ a constant that measures the intensity of correlation[7]). Although we will show numerical evaluations performed using the Gaussian random field model, our results are valid for any spatially correlated random processes, where correlation decreases with distance.

6.3.2.2 Spatio-Temporally Correlated Gaussian Random Field
Further, we consider spatio-temporal correlated processes $X(x, t)$, where x denotes the space dimension, and t denotes the time dimension. Our model for spatio-temporal processes is a generalization of the spatial model introduced in the previous section, with time considered as an additional dimension. We further assume that the process measured by the field is Gaussian distributed: each node measures a zero-mean and unit variance normal random variable $X(x, t) \approx \mathcal{N}(0, 1)$, which is

correlated both in space and time with the rest of the network nodes. We consider correlation structures of the form

$$E[X(x_1, t_1)X(x_2, t_2)] = \sigma_{X(x_1, t_1), X(x_2, t_2)}$$

$$= \sigma(|x_1 - x_2|, |t_1 - t_2|)$$

$$= e^{-c((x_1 - x_2)^2 + \gamma^2(t_1 - t_2)^2)\frac{\beta}{2}}, \qquad (6.1)$$

with γ as the scaling constant for the time axis.

6.4 Data Gathering of Spatially Correlated Random Processes

6.4.1 Lossless Data Gathering

Consider data gathering of random processes. The rate allocation at the nodes is denoted by $\{R_i\}_{i=1}^N$. Denote by ST an arbitrary spanning tree of G, and by c_i the total weight of the path connecting node i to designate sink S on the spanning tree ST. For a given network with a connectivity graph $G = (V, E)$, with V as the set the vertices, and with E as the set of edges connecting the vertices, we formulate our problem as follows (see Figure 6.1):

$$\{\{R_i^*, c_i^*\}_{i=1}^N, ST^*\} = \arg \min_{R_i, c_i, ST} \sum_{i \in V} R_i c_i \qquad (6.2)$$

under constraints

$$\sum_{i \in \mathcal{X}} R_i \geq H(\mathcal{X} | \mathcal{X}^C), \forall \mathcal{X} \subset V, \qquad (6.3)$$

where (6.3) provides the Slepian-Wolf constraints on rates for joint data representation at the nodes.

6.4.1.1 Optimal Transmission Structure Is the Shortest Path Tree (SPT)

The constraints in (6.3) imply that nodes can code with any rate that obeys the constraint region *without explicitly exchanging data*. As a consequence, we can state the following theorem.[9]

Theorem 6.1 *(Separation of the joint optimization of source coding and transmission structure)*

The overall joint optimization (6.2) can be achieved by first optimizing the transmission structure with respect only to the link weights c_i and then optimizing the rate allocation for the given transmission structure under the constraints (6.3).

Proof: By definition, for any given node, the cost function in (6.2) is separable as the product of a function that depends only on the rate allocated at that node and another function that depends only on the link weights. Once the rate allocation is *fixed*, the best way (least cost) to transport any amount of data from a given node i to the sink S does not depend on the value of the rate R_i. Since this holds for any rate allocation, it is also true for the minimizing rate allocation and the result follows.

Theorem 6.1 implies that the optimal transmission structure that optimizes (6.2) is the shortest path tree, given by the superposition of the shortest paths from all nodes to the sink $(ST^* = SPT)$.

Corollary 6.1 *(Optimality of the shortest path tree (SPT) for the single-sink data gathering problem)*
When there is a single sink S in the data gathering problem and Slepian-Wolf coding is used, the shortest path tree (SPT) rooted in S is optimal, in terms of minimizing (6.2), for any rate allocation.

Proof: The best way to transport the data from any node to the sink is to use the shortest path. Minimizing the sum of costs under constraints in (6.2) becomes equivalent to minimizing the cost corresponding to each node independently. Since the shortest path tree is a superposition of all the individual shortest paths corresponding to the different nodes, it is optimal for any rate allocation that does not depend on the transmission structure, which is the case here.

6.4.1.2 Rate Allocation

The total weight of the path from node i to the sink S denoted by $c_i^* = d_{SPT}(i, S)$. For the rest of this section, suppose without loss of generality that nodes are ordered in a list with increasing values of the weights corresponding to the shortest paths from each node to the sink, that is, $c_1^* \leq c_2^* \leq \cdots \leq c_N^*$.

From Corollary 6.1, it follows that the minimization of (6.2) becomes now a linear programming (LP) problem:

$$\{R_i^*\}_{i=1}^N = \arg \min_{\{R_i\}_{i=1}^N} \sum_{i \in V} R_i c_i^*, \tag{6.4}$$

under constraints (6.3).

Theorem 6.2   *(LP solution)*
   The solution of the optimization problem given by (6.4) under constraints (6.3) is[9].

$$R_1^* = H(X_1),$$
$$R_2^* = H(X_2|X_1),$$
$$\dots\dots\dots \tag{6.5}$$
$$R_N^* = H(X_N|X_{N-1}, X_{N-2}, \dots, X_1).$$

Theorem 6.2 is proven in Cristescu, Beferull-Lozano, and Vetterli.[10]
   Thus, for optimal rate allocation, nodes code by conditioning on all the other nodes that are closer to the sink on the SPT. In other words, the solution of this problem is given by the corner of the Slepian-Wolf region that intersects the cost function in exactly one point. The node with the smallest total weight on the SPT to the sink is coded with a rate equal to its unconditional entropy. Each of the other nodes is coded with a rate equal to its respective entropy conditioned on all other nodes which have a total smaller weight to the sink than itself.
   Figure 6.4 gives an example involving only two nodes, and it shows how the cost function is indeed minimized with such a rate allocation. Equation (6.5) corresponds in this particular case to the point $(R_1, R_2) = (H(X_1|X_2), H(X_2))$.
   Note that if two or more nodes are equally distanced from the sink on the SPT (e.g., $d_{SPT}(X_1, S) = d_{SPT}(X_2, S)$ in Figure 6.4), then the solution of (6.5) is not unique, since the cost function is parallel to one of the faces of the Slepian-Wolf region, so any point on the face is an optimal solution.
   For an illustration of Theorem 6.2, in Figure 6.5 we plot the typical rate allocation as a function of the node index for a one-dimensional network as in Figure 6.3 and a Gaussian process as introduced in Section 6.3.2.

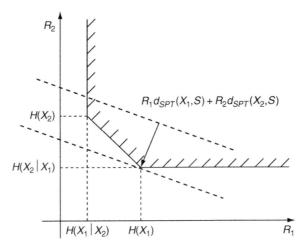

Figure 6.4 A simple example with two nodes. The total weights from sources X_1, X_2 to the sinks are, respectively, $d_{SPT}(X_1, S), d_{SPT}(X_2, S)$, and $d_{SPT}(X_1, S) < d_{SPT}(X_2, S)$ in this particular case. In order to achieve the minimization, the cost line $R_1 d_{SPT}(X_1, S) + R_2 d_{SPT}(X_2, S)$ has to be tangent to the most interior point of the Slepian-Wolf rate region, given by $(R_1, R_2) = (H(X_1), H(X_2|X_1))$.

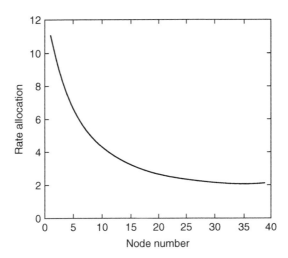

Figure 6.5 Typical rate allocation as a function of the node index for a one-dimensional network. The node index increases with the distance from the node to the sink.

Even if the solution can be provided in the closed form (6.5), a distributed implementation of the optimal algorithm at each node implies knowledge of the overall structure of the network (total weights between nodes and total weights from the nodes to the sink). This knowledge is needed for the following:

(a) Ordering the total weights on the SPT from the nodes to the sink: each node needs its index in the ordered sequence of nodes in order to determine on which other nodes to condition when computing its rate assignment. For instance, it may happen that the distance on the graph between nodes X_i and X_{i-1} is large. Thus, closeness in the ordering on the SPT does not mean necessarily proximity in distance on the graph.

(b) Computation of the rate assignment:

$$R_i = H(X_i|X_{i-1}, \ldots, X_1) = H(X_1, \ldots, X_i) - H(X_1, \ldots, X_{i-1}).$$

Note that for each node i we need to know locally *all* distances among the nodes $X_1, \ldots, X_i$, $i > 1$, in order to be able to compute this rate assignment, because the rate assignment involves a conditional entropy including all these nodes. However, for a static network, these distances can be calculated off-line at the deployment of the network. As a result, the optimal rate allocation can be computed at the beginning of the operation as well.

This implies that, for a distributed algorithm, global knowledge should be available at nodes, which might not be the case in a practical situation.

However, note that if the correlation decreases with distance, as is usual in sensor networks, it is intuitive that each node i could condition only on a small neighborhood, incurring only a small penalty. In the next subsection, we propose a fully distributed heuristic approximation algorithm, which avoids the need for each node to have global knowledge of the network and provides solutions for the rate allocation which are very close to the optimum.

6.4.1.3   Heuristic Approximation Algorithm

So far, we have found the optimal solution of the LP problem for the rate assignment under the Slepian-Wolf constraints. In this subsection, we consider the design of a distributed heuristic approximation algorithm for the case of single-sink data gathering.

Suppose each node i has complete information (distances between nodes and total weights to the sink) only about a local vicinity $\mathcal{N}_1(i)$ formed by its immediate neighbors on the connectivity graph G. All this information can be computed in a distributed manner by running, for example, a distributed algorithm for finding the SPT (e.g., Bellman-Ford[5]). By allowing a higher degree of (local) overhead communication, it is also possible for each node i to learn this information for a neighborhood $\mathcal{N}_k(i)$ of k-hop neighbors. The approximation Algorithm 6.1 that we propose is based on the observation that nodes that are outside this neighborhood count very little, in terms of rate, in the local entropy conditioning.

Algorithm 6.1 Approximated Slepian-Wolf coding

---

for each node i **do**
    Set the neighborhood range k (only k-hop neighbors).
    Find the *SPT* using a distributed Bellman-Ford algorithm.
end for
for each node i **do**
    Using local communication, obtain all the information from the
        neighborhood $\mathcal{N}_k(i)$ of node i.
    Find the set $\mathcal{C}_i$ of nodes in the neighborhood $\mathcal{N}_k(i)$ that are closer
        to the sink, on the *SPT*, than the node i.
    Transmit at rate $R_i^\dagger = H(X_i|X_j, j \in \mathcal{C}_i)$.
end for

This means that data are coded locally at the node with a rate equal to the conditional entropy, where the conditioning is performed *only* on the subset $\mathcal{C}_i$ formed by the neighbor nodes which are closer to the sink than the respective node.

The proposed algorithm needs only local information, so it is completely distributed. For a given correlation model, depending on the reachable neighborhood range, this algorithm gives a solution close to the optimum since the neglected conditioning is small in terms of rate for a correlation function that decays sufficiently fast with the distance.

We present numerical simulations that show the performance of this approximation algorithm for the case of single-sink data gathering. We consider the stochastic data model introduced in Section 6.3.2, given by a multivariate Gaussian random field, and a correlation

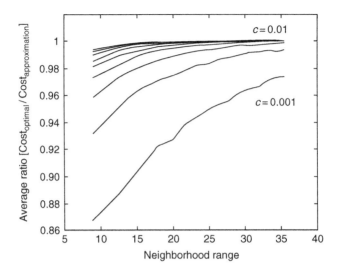

Figure 6.6 Slepian-Wolf coding: average value of the ratio between the optimal and the approximated solution, in terms of total cost, vs. the neighborhood range. Every network instance has 50 nodes uniformly distributed on a square area of size 100×100, and the correlation exponent varies from $c = 0.001$ (high correlation) to $c = 0.01$ (low correlation). The average has been computed over 20 instances for each (c, radius) value pair.

model where the internode correlation decays exponentially with the distance between the nodes. More specifically, we use an exponential model of the covariance $K_{ij} = \exp(-cd_{i,j}^2)$. The weight of an edge (i, j) is $w_{i,j} = d_{i,j}^2$, and the total cost is given by (6.4). Figure 6.6 presents the average ratio of total costs between the Slepian-Wolf approximated solution, using a neighborhood of $\mathcal{N}_1(i)$ for each node, and the optimal one. In Figure 6.7, we show a comparison of our different approaches for the rate allocation, as a function of the distances from the nodes to the sink.**

6.4.1.4 Scaling Laws: A Comparison between Slepian-Wolf Coding and Explicit Communication-Based Coding
The alternative to Slepian-Wolf coding is coding by explicit communication, which is considered by Cristescu and Colleagues.[12,14] In this

---

** Explicit communication is introduced in Section 6.4.1.4.

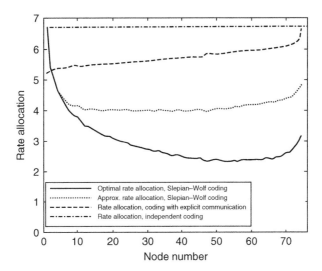

Figure 6.7 Average rate allocation for 1000 network instances with 75 nodes and a correlation exponent $c = 0.0008$ (strong correlation). On the x-axis, nodes are numbered in increasing order as the total weight from the sink increases on the corresponding SPT.

case, compression at nodes is done only using *explicit communication* among nodes, namely, a node can reduce its rate only when data from other nodes that use it as relay are available (as opposed to Slepian-Wolf coding, where no communication among nodes is required for joint optimal rate allocation). The study of the complexity of joint rate allocation and transmission structure optimization with explicit communication can be found in Cristescu et al.[12,14]

In this subsection, we compare the asymptotic behavior (large networks) of the total cost using Slepian-Wolf coding and the total cost with coding by explicit communication. The advantages that coding by explicit communication has over Slepian-Wolf coding are (i) no a priori knowledge of the correlation structure is needed and (ii) the compression, which is done by conditional encoding, is easily performed at the nodes relaying data. However, even for a simple one-dimensional setting presented in this section, our analysis shows that in large networks, for some cases of correlation models and network scalability, Slepian-Wolf coding can provide very important gains over coding by explicit communication in terms of total flow cost.

For the sake of simplicity in the analysis, we consider a one-dimensional network model where there are N nodes placed uniformly

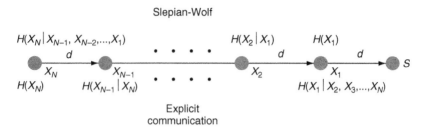

Figure 6.8 A one-dimensional example: the rate allocations for Slepian-Wolf (above the line) and explicit communication (below the line).

on a line (see Figure 6.8). The distance between two consecutive nodes is d. The nodes need to send their correlated data to the sink S.

For this scenario, the SPT is clearly the optimal data gathering structure for both coding approaches. Thus, the overall optimization problem (6.2) simplifies, and we can compare the two different rate allocation strategies in terms of how they influence the total cost.

Within the one-dimensional model, we consider two important cases of network scalability, namely, the *expanding network*, where the internode distance is kept constant and equal to $d = 1$ (i.e., by increasing N, we increase the distance between the node N and the sink S), and the *refinement network*, where the total distance from node N to the sink S is kept constant, namely, $Nd = 1$ (that is, nodes are uniformly placed on a line of length one, and hence, by adding nodes, the internode distance goes to zero).

As mentioned in Section 6.3.2, we consider that the nodes of the network are sampling a Gaussian continuous-space wide-sense-stationary (WSS) random process $X_c(s)$, where s denotes the position. Thus, we have a vector of correlated sources $\mathbf{X} = (X_1, \ldots, X_N)$, where $X_i = X_c(id)$, and the correlation structure for the vector $\mathbf{X}$ is inherited from the correlation present in the original process $X_c(s)$. As N goes to infinity, the set of correlated sources represents a discrete-space random process denoted by $X_d(i)$, with the index set given by the node positions. Thus, the spatial data vector $\mathbf{X}$ measured at the nodes has an N-dimensional multivariate normal distribution $G_N(\mu, \mathbf{K})$. In particular, we consider two classes of random processes:

(a) Non-bandlimited processes, namely, (1) $K_{ij} = \sigma_{ij}^2 \exp(-c|d_{i,j}|)$, which corresponds to a regular continuous-space process,[23] and (2) $K_{ij} = \sigma_{ij}^2 \exp(-c|d_{i,j}|^2)$, which corresponds to a singular continuous-space process,[23] where $c > 0$.

(b) Bandlimited process with bandwidth B, that is, there exists a continuous angular frequency such that $S_{X_c}(\Omega) = 0$, for $|\Omega| \geq \Omega_0$, where $S_{X_c}(\Omega)$ is the spectral density and $B = 2\Omega_0$. This process can also be shown to be a singular continuous-space process.[23][†]

Let us denote the conditional entropies by $a_i = H(X_i|X_{i-1}, \ldots, X_1)$. Note that for any correlation structure, the sequence a_i is monotonically decreasing (because conditioning cannot increase entropy) and is bounded from below by zero (because the entropy cannot be negative). Since the nodes are equally spaced, and the correlation function of a WSS process is symmetric, it is clear that $H(X_I|X_{I-1}, X_{I-2}, \ldots, X_{I-i}) = H(X_I|X_{I+1}, X_{I+2}, \ldots, X_{I+i})$, for any $I, 0 \leq i \leq I - 1$.

Let us denote by $\psi(N)$ the ratio between the total cost associated to Slepian-Wolf coding ($\text{cost}_{SW}(N)$) and the total cost corresponding to coding by explicit communication ($\text{cost}_{EC}(N)$), that is,

$$\psi(N) = \frac{\text{cost}_{SW}(N)}{\text{cost}_{EC}(N)} = \frac{\sum_{i=1}^{N} i a_i}{\sum_{i=1}^{N}(N - i + 1)a_i}. \tag{6.6}$$

Then, the following theorem holds.[11]

Theorem 6.3   *(Scaling Laws)*
  Asymptotically, we have the following results:

(i) *If* $\lim_{i \to \infty} a_i = C > 0$,
   - $\lim_{N \to \infty} \psi(N) = 1$,
   - $\text{cost}_{SW}(N) = \Theta[\text{cost}_{EC}(N)]$.

(ii) *If* $\lim_{i \to \infty} a_i = 0$,

   (ii)-1 *If* $a_i = \Theta(1/i^p), p \in (0, 1)$,
      - $\lim_{N \to \infty} \psi(N) = 1 - p$,
      - $\text{cost}_{SW}(N) = \Theta[\text{cost}_{EC}(N)]$.

[†] Actually, it can be shown that the same singularity property holds as long as $S_{X_c}(\Omega) = 0$ on some frequency interval of non-zero measure.[23]

(ii)-2 If $a_i = \Theta(1/i^p), p \geq 1,$
- $\lim_{N \to \infty} \psi(N) = 0,$
- $cost_{SW}(N) = o[cost_{EC}(N)],$
- If $p = 1, \psi(N) = \Theta(1/\log N),$
- If $p \in (1,2), \psi(N) = \Theta(1/N^{p-1}),$
- If $p = 2, \psi(N) = \Theta(\log N/N),$
- If $p > 2, \psi(N) = \Theta(1/N).$

The proof of Theorem 6.3 can be found in Cristescu, Beferull-Lozano, and Vetterli. In Figure 6.9, we show typical behaviors of the ratio of total flow costs for the two coding approaches.

   We now apply Theorem 6.3 to the correlation models we consider in this chapter.

- For an expanding network: In cases (a.1) and (a.2), the result of sampling is a discrete-space regular process,[29] thus $\lim_{i \to \infty} a_i = C > 0$, and it follows that $\lim_{N \to \infty} \psi(N) = 1$. In case (b), if the spatial sampling period d is smaller than the Nyquist sampling rate

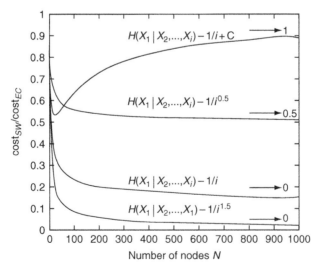

Figure 6.9   Typical behavior of the ratio of the total costs $cost_{SW}(N)/cost_{EC}(N)$.

$1/B$ of the corresponding original continuous-space process, then $\lim_{i\to\infty} a_i = 0$. The specific speed of convergence of a_i depends on the spatial sampling period (that is, how small it is with respect to $1/B$) and the specific (bandlimited) power-spectrum density function of the process. In Figure 6.10, we show the bandlimited example with correlation $K(\tau) = B\mathrm{sinc}(B\tau)$. It can be seen that when $d < 1/B$, $a_i = o(1/i)$, and, thus, the ratio of total costs goes to zero. Also, the smaller d is, the faster the convergence is.

- For a refinement network: In case (a.1), it is shown[10] that $H(X_i|X_{i-1},\ldots,X_1) = H(X_i|X_{i-1})$, thus $a_i = H(X_i|X_{i-1})$ for any $i \geq 2$. Then, for any finite N, $a_N > 0$. Since ia_i does not converge to zero (see Figure 6.11), then it follows from Theorem 6.3 that, in the limit, the ratio of total costs is $\lim_{N\to\infty} \psi(N) = 1$. In case (a.2), a closed form expression for the conditional entropy is difficult to derive. However, we show numerically in Figure 6.11(a) that in this case a_i decreases faster[‡] than $1/i$. Thus, from Theorem 6.3, $\lim_{N\to\infty} \psi(N) = 0$. For comparison purposes, we show in Figure 6.11(a) the behavior for case (a.1). In Figure 6.11(b), we plot also the ratio of total costs

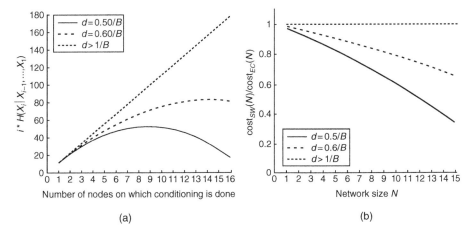

(a)   (b)

Figure 6.10 Expanding network sampling a bandlimited process with correlation model given by $K(\tau) = B\mathrm{sinc}(B\tau)$. (a) The conditional entropy $H(X_i|X_{i-1},\ldots,X_1)$ decreases faster than $1/i$ if $d < 1/B$. (b) The behavior of the ratio of total $\mathrm{cost}_{SW}(N)/\mathrm{cost}_{EC}(N)$ as a function of the size of the network.

[‡] Since these two processes are both non-bandlimited, sampling them results in discrete-space regular processes.[29] However, the sampled model (a.2) inherits a "superior predictability" than (a.1), which makes a_i decrease faster than $1/i$.

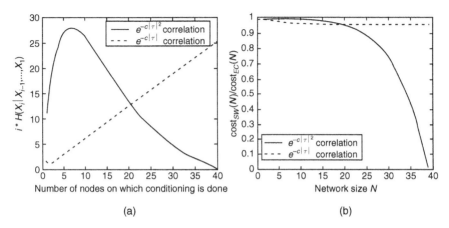

Figure 6.11 Correlation dependence on the internode distance d given by $\exp(-c|\tau|^{\beta}), \beta \in \{1, 2\}$. (a) The conditional entropy $H(X_i|X_{i-1}, \ldots, X_1)$ decreases faster than $1/i$ for $\beta = 2$, but is constant for $\beta = 1$ (after $i \geq 2$). (b) The behavior of the ratio $\text{cost}_{SW}(N)/\text{cost}_{EC}(N)$ as a function of the size of the network.

for both correlation models. Finally, in case (b), a_i goes to zero very fast, as for the case (a.2), because of the singularity of the original bandlimited process. It can be seen in Figure 6.12 how the ratio of costs starts to decrease as soon as $d < 1/B$, thus $\lim_{N \to \infty} \psi(N) = 0$.

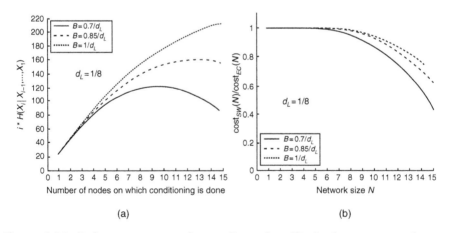

Figure 6.12 Refinement network sampling a bandlimited process; we denote the reference bandwidth with $B_L = 1/d_L$. (a) The conditional entropy $H(X_i|X_{i-1}, \ldots, X_1)$ decreases faster than $1/i$ as soon as $B < 1/d$, that is, $N > 1/d_L$. (b) The behavior of the ratio of total $\text{cost}_{SW}(N)/\text{cost}_{EC}(N)$ as a function of the size of the network.

Intuitively, results similar to those presented in this section hold also for higher dimensions, when the transmission structure that is used is the same (e.g., SPT) for both types of coding. The ideas leading to the results for the one-dimensional network can be generalized to two-dimensional networks. For instance, one can consider a two-dimensional wheel structure with the sink in the center of the wheel, where entropy conditioning at the nodes on any spoke is done as in the one-dimensional case (see Figure 6.13). The same analysis as in the one-dimensional case holds, with the additional twist that, according to Theorem 6.2, Slepian-Wolf coding at node i is done by conditioning not only on the nodes closer to the sink on its spoke, but also on the nodes on the other spokes closer to the sink on the SPT than node i (the dashed circle in Figure 6.13). However, the explicit communication coding is still done only on the nodes on the spoke that forward their data to node i (the solid circle in Figure 6.13). Thus, the ratio of costs $\psi(N)$ in the two-dimensional case is upper bounded by its counterpart in the one-dimensional case, which means that the results of Theorem 6.3 apply for the two-dimensional case as well.

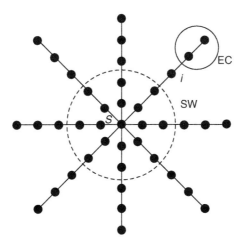

Figure 6.13   A two-dimensional network with a wheel structure, with the sink S in the center. Slepian-Wolf coding for node i is done by conditioning on the nodes in the dashed region (denoted by SW). Explicit communication coding for node i is done by conditioning on nodes in the solid region (denoted by EC).

6.4.2 Node Placement

Further, we consider the related problem where a given number of nodes N is placed in a field such that the sensed data can be reconstructed at the sink within specified distortion bounds while minimizing the energy consumed for communication. For the sake of simplicity, consider the one-dimensional network in Figure 6.3, where the nodes can now be placed in arbitrary position on the line. Denote by w_i the distance between node i and node $i-1$. The optimization problem considered in this section aims to minimize the total transmission cost under distortion constraints. The distortion constraints impose a prescribed accuracy of representation at the sink for all the points in the network, when only the measurements at the sensor nodes are available.

Note that for the one-dimensional network in Figure 6.3 and for a space-dependent correlation model as considered in this chapter, the distortion constraints translate into spatial constraints. Assume the value at intermediate nodes is approximated by the sensing node corresponding to the Voronoi cell to which that intermediate node belongs. Then, our optimization becomes

$$\{R_i^*, w_i^*\}_{i=1}^N = \arg\min_{R_i, w_i} \sum_{i=1}^N \sum_{j=i}^N R_j \times w_i^\kappa \tag{6.7}$$

under constraints

$$\sum_{i=1}^{N-1} w_i + \frac{w_N}{2} = L, \tag{6.8}$$

$$w_i \le W^{\max}, \tag{6.9}$$

where (6.8) is a constraint on the average distortion, and (6.9) is a constraint on the individual constraints per Voronoi cell.

For the sake of simplicity, we only consider in this chapter the case where data at nodes are independent, namely, $\{R_i\}_{i=1}^N = R$. A complete analysis of the case when Slepian-Wolf coding and explicit communication-based coding are exploited is provided in Ganesan, Cristescu, and Beferull-Lozano.[16]

Then, the optimal placement $\{w_i^*\}_{i=1}^N$ is obtained by solving

$$\{w_i^*\}_{i=1}^N = \arg\min_{w_i} \sum_{i=1}^N (N-i+1) w_i^\kappa, \tag{6.10}$$

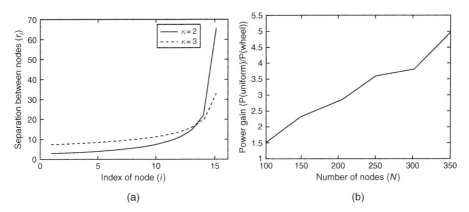

Figure 6.14 Optimal placement for a one-dimensional network for different path-loss exponents (left); and energy improvement over uniform placement for the two-dimensional network (right) as a function of the number of network nodes.

under the distortion constraints (6.8) and (6.9).

By using Lagrangian optimization with a multiplier λ, we obtain the optimal solution

$$w_i = \left(\frac{\lambda}{\kappa(N-i+1)}\right)^{\frac{1}{\kappa-1}}, i = 1 \ldots N-1,$$

$$w_N = \left(\frac{3\lambda}{2\kappa}\right)^{\frac{1}{\kappa-1}},$$

$$\lambda = \left(\frac{L}{\sum_{i=1}^{N-1}\left(\frac{1}{\kappa(N-i+1)}\right)^{\frac{1}{\kappa-1}} + \frac{3}{2}\left(\frac{3}{2\kappa}\right)^{\frac{1}{\kappa-1}}}\right)^{\kappa-1}.$$

Such a placement provides important energy performance improvements as compared to uniform placement (see Figure 6.14), and exploiting data correlation by using Slepian-Wolf coding further improves the results.[16]

6.4.3 Lossy Data Gathering

6.4.3.1 Rate-Distortion Allocation

Let us further consider the case when lossy coding is used at nodes, but with high resolution.[37] The information measured by the nodes should be available at the sink within certain total and individual distortion bounds. A rate-distortion allocation $\{(R_i, D_i)\}_{i=1}^N$ (bits) has to be assigned at the nodes so that the quantized measured information samples are described with certain total D and individual $D_i^{\max}, i = 1, \ldots, N$ distortions. The spanning tree to be found is denoted by ST, which defines the transmission structure; $c_i, i = 1 \ldots N$ are the total weights of the path from node i to the sink on the spanning tree ST, thus $c_i = \sum_{e \in \mathcal{E}_i} w_e^\kappa$, where $e \in \mathcal{E}_i$, $\mathcal{E}_i$ is the set of edges linking node i to the sink S on ST, w_e is the Euclidean distance of edge e, and κ is the path-loss exponent. $h(\cdot)$ denotes the differential entropy. Then, the most general form of our optimization problem is given as follows:

$$\{\{R_i^*, D_i^*, c_i^*\}_{i=1}^N, ST^*\} = \arg \min_{R_i, D_i, c_i, ST} \sum_{i=1}^N c_i R_i \qquad (6.11)$$

under constraints

$$\sum_{i \in \mathcal{X}} R_i \geq h(\mathcal{X} | V \backslash \mathcal{X}) - \log \prod_{i \in \mathcal{X}} 2 \pi e D_i, \forall \mathcal{X} \subset V, \qquad (6.12)$$

$$\sum_{i=1}^N D_i \leq D, \quad D_i \leq D_i^{\max}, i = 1 \ldots N, \qquad (6.13)$$

where (6.12) expresses the rate-distortion region constraints given in Zamir and Berger,[37] namely, that the sum of rates for any given subset of nodes has to be larger than the entropy of the random variables measured at those nodes, conditioned on the random variables measured at all other nodes. In the constraints in (6.13), D is the maximum allowed total distortion, and $D_i^{\max}, i = 1 \ldots N$ are the maximum individual constraints.

In the high rate regime, uniform quantization and Slepian-Wolf coding is optimal.[37] Thus, a similar result as the one stated in Theorem 6.1 holds in the lossy case about the separation between transmission optimization and rate-distortion allocation, since nodes do not need to communicate explicitly to code data with a given rate-distortion allocation. As a result, the SPT is the optimal transmission structure in this case as well. Thus, we can prove in the lossy case a result similar to Theorem 6.2 which corresponds to the lossless case; namely, for any set of distortion values $\{D_i\}_{i=1}^N$, the rate allocation is given by Theorem 6.4.

Theorem 6.4 *(Optimal rate allocation):*

$$R_1^* = h(X_1) - \log 2\pi e D_1,$$
$$R_2^* = h(X_2|X_1) - \log 2\pi e D_2,$$

$$\cdots \tag{6.14}$$

$$R_N^* = h(X_N|X_{N-1}, \ldots, X_1) - \log 2\pi e D_N.$$

Theorem 6.4 is proven in Cristescu and Beferull-Lozano.[8]

Next, we consider optimization of (6.11) for the case where we assume that the individual distortion constraints given by (6.13) are not active. By Theorem 6.4, $\{R_i^*\}_{i=1}^N$ *only* depends on $\{D_i\}_{i=1}^N$. Therefore, at this point, we can insert in (6.11) the values for $\{R_i^*\}_{i=1}^N$ given by Theorem 6.4, thus obtaining an optimization problem having as argument the set of distortions $\{D_i\}_{i=1}^N$ only, that is,

$$\{D_i^*\}_{i=1}^N = \arg \min_{\{D_i\}_{i=1}^N} \sum_{i=1}^N c_i^* \cdot (h(X_i|X_{i-1}, \ldots, X_1) - \log 2\pi e D_i)$$

under the constraint

$$\sum_{i=1}^N D_i \leq D. \tag{6.15}$$

Note that the differential entropy terms in (6.15) do not depend on the distortions D_i. Thus, it can be equivalently written as

$$\{D_i^*\}_{i=1}^N = \arg \max_{\{D_i\}_{i=1}^N} \sum_{i=1}^N c_i^* \log 2\pi e D_i$$

under the constraint

$$\sum_{i=1}^N D_i \leq D. \tag{6.16}$$

Denote by $\sum_{i=1}^N c_i^* = C$. The solution of the optimization problem (6.16) is easily obtained, using Lagrange multipliers, giving a linear distribution of distortions:

$$D_i^* = D \cdot \frac{c_i^*}{C}, \quad i = 1 \ldots N. \tag{6.17}$$

By combining (6.14) and (6.17), the rate-distortion allocation at nodes is given by

$$R_i^* = h(X_i|X_{i-1}, \ldots, X_1) - \log\left(\frac{2\pi e D c_i^*}{C}\right), \quad i = 1 \ldots N. \tag{6.18}$$

Note that, for correlation functions that depend only on the distance among nodes, as we consider in this chapter, and for a uniform placement of nodes, the differential entropy monotonically decreases as the number of nodes on which conditioning is done increases. Also, by definition, the sequence $\{c_i^*\}_{i=1}^N$ is monotonically increasing with i. As a result, the rate allocation R_i in (6.18) is a function that monotonically decreases with the node index i.

Further, (6.18) essentially depends only on the path weights ordering of the nodes on the SPT, given by $\{c_i^*\}_{i=1}^N$. For example, in the case of correlated Gaussian random fields, (6.18) can be written as

$$R_i^* = \log\frac{\det(\mathbf{K}(1,\ldots i))}{\det(\mathbf{K}(1,\ldots,i-1))}\frac{C}{c_i^* \cdot D}, \quad i = 1 \ldots N, \tag{6.19}$$

where $\mathbf{K}(1, \ldots, i)$ is the correlation matrix corresponding to nodes $1, \ldots, i$.

In Figure 6.15, we illustrate the distortion and rate allocations provided by the result in (6.19) for the one-dimensional grid with uniform internode distances, measuring a correlated Gaussian random field with $\beta = 1, c = 10^{-3}, N = 20, D = 10^{-3}, D_i^{max} = 0.7 \cdot 10^{-3}$, and

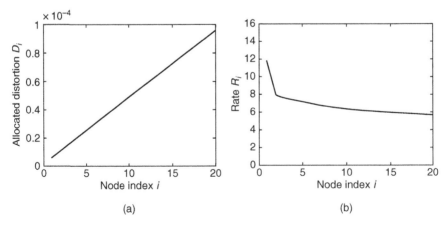

Figure 6.15 One-dimensional network average distortion constraint: (a) distortion and (b) rate allocations as a function of the node index.

$\kappa = 2$ (which corresponds to sampling a continuous Gauss-Markov process). The same analysis holds for arbitrary two-dimensional networks. Similar results are obtained for many other network parameter settings, and in the case of individual distortion constraints.[8]

6.4.3.2   Node Placement

In some scenarios, the positions of the nodes are not fixed in advance, but it is possible to place the nodes optimally so as to minimize various resources.[16] Since the study in this work is concerned with energy efficient scenarios, one possible task to be achieved, when the node placement can be chosen, is the total energy efficiency. We consider the optimization (6.11) where the positions of the nodes are additional variables, and the optimization is done additionally over the weights of the paths from the nodes to the sink:

$$\{R_i^*, D_i^*, c_i^*\}_{i=1}^N = \arg \min_{\{R_i, D_i, c_i\}_{i=1}^N} \sum_{i=1}^N c_i R_i \qquad (6.20)$$

under the constraints specified by (6.12) and (6.13), with additional constraints on $\max\{c_i\}_{i=1}^N$ (coverage constraint) and on $\max\{c_i - c_{i-1}\}_{i=2}^N$ (internode space constraint). The coverage constraint imposes that the entire area is covered. The internode space constraint ensures that any given point in the measured area is close enough to a sensor node, such that the data corresponding to that point can be reconstructed with a certain minimal accuracy at the sink by approximating it with the value measured by the closest sensor.

We study the problem of optimal placement for two energy efficiency targets of interests, namely, total energy and network lifetime,[§] and compare the tradeoffs involved. For the one-dimensional example in Figure 6.3, the optimal placement is[8]

$$w_i^* = \frac{L}{(\sum_{j=i}^N R_j^*)^\delta \left(\sum_{l=1}^N \frac{1}{(\sum_{j=l}^N R_j^*)^\delta} \right)}, i = 1 \ldots N,$$

where $\{w_i^*\}_{i=1}^N$ is the distance between nodes $i-1$ and i, $\delta = 1$ for total energy minimization, and $\delta = 1/2$ for lifetime maximization.

[§] We address network lifetime optimization by considering the constraint that all nodes use equal energy.

The optimal joint solution for the placement and rate allocation is obtained by using an iterative algorithm (Algorithm 6.2).

Algorithm 6.2 Placement and rate allocation

Initialize uniformly the node placement $\{w_i\}_{i=1}^N = L/N$.
repeat
    Given $\{w_i\}_{i=1}^N$, solve the energy minimization problem for $\{R_j\}_{j=1}^N$.
    Rewrite $\{w_i\}_{i=1}^N$ as a function of $\{R_j\}_{j=1}^N$.
until convergence.

The energy consumption and node positioning resulting from running Algorithm 6.2 are presented in Figures 6.16 (for the case of total energy minimization) and 6.17 (for the case of lifetime maximization). A complete study of the node placement problem for efficient lossy network data gathering is provided in Cristescu and Beferull-Lozano.[8]

6.4.4 Remarks

In this section, the problem of data gathering from a network of nodes measuring correlated data to a sink node was studied. An

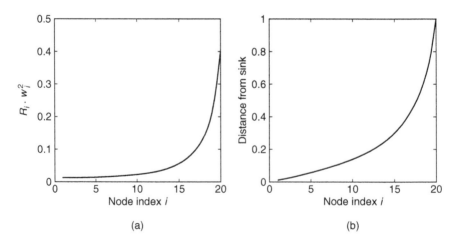

(a)   (b)

Figure 6.16 Placement optimization for minimizing the total energy: (a) energy consumption as a function of node index, and (b) distance from sink as a function of node index.

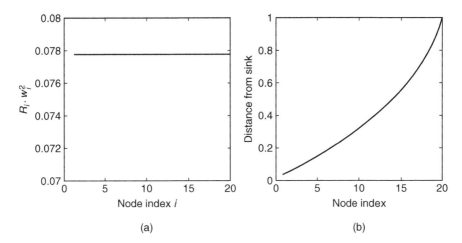

Figure 6.17 Placement for lifetime optimization: (a) energy consumption as a function of node index, and (b) distance from sink as a function of node index.

energy-related metric was considered, and we showed that in both the cases of lossless and lossy coding of the measured information, the optimal transmission structure is the SPT of the network graph. Further, the optimal rate and distortion allocations for coding at nodes were determined. Moreover, for the case when the positions of the nodes can be chosen, we provided strategies for energy efficient node placement.

Our discussion so far considers the case of spatially correlated but temporally i.i.d. processes. In the following section, we will discuss the case of processes that are correlated both in time and in space.

6.5 Data Gathering of Spatio-Temporal Processes

In this section, we consider the usual scenario of a sensor network with a sink. The goal of the sink is now to reconstruct the *entire* field over space and time, with a certain minimum distortion and only based on measurements at the sensors.[13] For this, a fixed number of nodes N are placed in the field (see Figure 6.18). Nodes transmit their measurements to the sink, at given time instants, by using a subset of the links of the graph. The sink needs to reconstruct the whole field with a minimal total distortion. We consider settings where data are time critical, and thus, delay results in distortion. Such settings

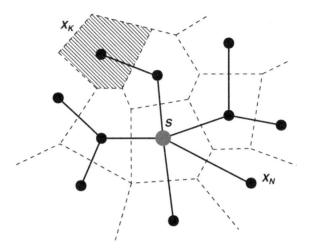

Figure 6.18 In this example, spatio-temporally correlated data from nodes $1, \ldots, N$ need to arrive at sink S. The sink needs to reconstruct the *whole* data field using only the measured values $X_1, \ldots, X_N$. Arbitrary points in the two-dimensional space are approximated by the measurements of the sensor node corresponding to the Voronoi cell to which they belong. The dashed zone corresponding to node k represents the area of the field approximated by the value X_k in node k. The thick solid lines show a chosen transmission structure (here, the SPT). Data from node k reaches the sink after being relayed by one other sensor node.

include scenarios for fire prevention, seismic awareness, and sensor networks measuring phenomena where abrupt transitions are critical (e.g., cracks in a massive structure or mudslides over a large terrain area). Another class of relevant scenarios is when sink feedback or control is needed at nodes and where the effect of delay in reporting the data induces suboptimal estimation and control.

If the network is dense (large N), the data have a good spatial approximation. However, for energy efficiency reasons, nodes in sensor networks cannot transmit their data directly to the sink, but rather communication is usually done via energy efficient transmitting structures. This implies long delays until the data sent from nodes far from the sink reach the sink, which thus results in weak temporal approximation.

On the contrary, the opposite effects take place when N is small. The spatial approximation of the data is poor, but on the other hand data have to travel over only a limited number of hops, which results in a good temporal approximation. Thus, as we will show in this section, for a given spatio-temporal correlation structure, there usually exists a *finite optimal N* that minimizes the overall distortion of the field reconstruction at the sink.

6.5.1 Problem Setup

We study the influence of node density on the total distortion of estimation when several aspects specific to sensor nodes are considered, namely, delay and energy efficiency. Our setting takes into account two important issues typical in sensor networks scenarios: the precision of estimation[22] and the energy efficiency.[30]

First, since the measured data are correlated and the number of available nodes is limited, the sink can reconstruct the values of the field at each point by approximating them with the values at the points where the actual sensor nodes are placed. In those points, full measurements are available. Also, no other information except the values measured at sensor nodes is available at the sink about that region of the field. The precision of the approximation depends both on the level of spatial correlation in the data and on the number of sensors available. This approximation introduces a first factor of distortion, which we call "spatial distortion."

Second, since the nodes have limited battery energy, a good strategy is to send data via relaying nodes rather than directly to the sink (multi-hopping). However, multi-hopping results in data delay, since data from the extremities of the network need to be transmitted via multiple relays until they reach the sink. In various practical sensor network scenarios, data are needed at the sink in real-time. For instance, for the tasks of control or active monitoring, data may become useless if it arrives at the sink with too large a delay. For a spatio-temporal correlated process, the data that arrive at the sink are distorted from the original measured values; however, they can be reconstructed with a certain precision given by the intensity of temporal correlation of the process. Thus, delay introduces a second factor of distortion, which we call "time distortion."

In this chapter, the two types of distortion are modelled as a single *distortion per field point* quantity. Their combined effect results in the total distortion of the field at the sink, and the goal of this chapter is to study how this quantity is influenced by the density of nodes of the sensor network. We argue that for various typical spatio-temporal correlation models of the data field, there is a unique optimal value for the number of placed nodes N that minimizes the total spatio-temporal distortion.

6.5.2   One-Dimensional Networks

6.5.2.1   Transmission Model

For simplicity of the analysis, in this chapter we consider networks with nodes uniformly placed on one- and two-dimensional grids, for which the distortion optimization is done only with respect to the size of the network. In the case of arbitrary networks with position-dependent correlation structure, the optimization becomes a function of the node placement as well. In this chapter we consider Gaussian random processes which exhibit spatio-temporal correlation (introduced in Section 6.3.2).

Consider N nodes placed on a line of fixed length L (see Figure 6.19). The internode distance is $d = L/N$. An additional node S at the extreme right of the line represents the sink to which all data should arrive. The task of the sink is to reconstruct the *whole* field on the line.

We assume that the quantization done at nodes is very fine, namely, we assume the reconstruction error at the sink is only an estimation error. For that, points on the line that belong to the space intervals among the N nodes are assigned to Voronoi cells of the sensor nodes; these cells are delimited by mid-interval points. Therefore, each sensor node covers an interval of length d around its position. The values of the intermediate points are estimated at the sink by the value of

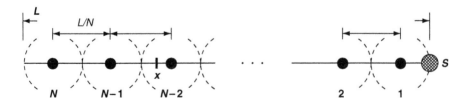

Figure 6.19   A one-dimensional example of a data gathering network, where each sensor node covers an area of the whole network. An *instantaneous point of reconstruction* is denoted by x. In this example, x belongs to the Voronoi cell of node $N - 2$. Thus, its value at time t is approximated, with a certain distance-dependent distortion, by $X(x_{N-2}, t)$.

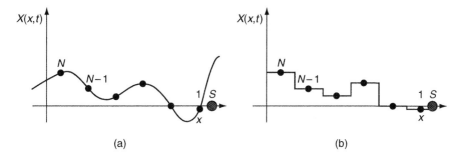

Figure 6.20 The reconstruction at the sink is based on the values measured by the sensing nodes: (a) the original signal, and (b) the approximation by interpolation at the sink.

the corresponding sensor node in the middle of the Voronoi cell[*] (see Figures 6.19 and 6.20). We assume that the measurements at nodes have the same variance and that the coding rates allocated at nodes are equal. This implies that nodes use equal transmission energy and that the spatial distortion per cell does not depend on the node identity.

   Multi-hopping, as mentioned in Section 6.5.1, inherently introduces delay. The delay has two causes. On the one hand, if relaying nodes have finite buffers, then packet forwarding results in a certain processing time due to buffering, and thus, the delay is proportional to the number of hops that data need to travel. On the other hand, if communication edges are lossy channels, then this requires retransmissions of data, and thus, the delay is proportional to the number of edges that data need to travel. For the sake of simplicity, we assume that the delay is proportional to the number of hops. Therefore, since the measured process is correlated both in time and in space, the reconstruction is further distorted from the real-time value due to delay caused by relaying.

6.5.2.2   Point-Wise Distortion

For instance, consider the value $X(x, t_0)$ of an arbitrary point x on the line at an arbitrary time t_0 (see Figure 6.21). Assume that the sink approximates the value at point x by considering the value $X(x_0, t_0)$ at point x_0 placed k hops away from the sink. For any data packet, we assume that the relation between the time delay t_k of that packet and

---

[*] More complex estimation strategies can be designed, in which the field value at an arbitrary position is based on the measurements of more than a single sensor node.[25,26]

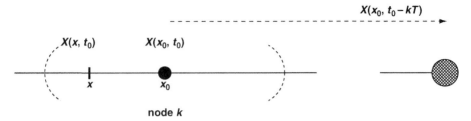

Figure 6.21 The value at point x is approximated by the value of the kth sensor node placed at x_0, resulting in *spatial distortion*. Moreover, due to transmission over k hops, the version that reaches the sink is delayed with kT, which results in *time distortion*. The combined result of the two distortion effects is the *total distortion* $D_x(N)$.

the number of hops k it has to travel is $k = \gamma t_k$. Assume the delay per hop is a constant T. Thus, $t_k = kT$. Then, the MSE of $X(x, t_0)$ at the sink, when $X(x_0, t_0)$ is known, is expressed by

$$
\begin{aligned}
D_{x,t_0,x_0,k} &= E\left[(X(x,t_0) - X(x_0,t_0+kT))^2\right] \\
&= E\left[X(x,t_0)^2\right] + E\left[X(x_0,t_0+kT)^2\right] - 2E\left[X(x,t_0)X(x_0,t_0+kT)\right] \\
&= 2 - 2\sigma_{X(x,t_0)X(x_0,t_0+kT)} \\
&= 2 - 2\sigma(|x - x_0|, kT) \\
&= 2 - 2e^{-c((x-x_0)^2 + (\gamma kT)^2)^{\frac{\beta}{2}}}.
\end{aligned}
\tag{6.21}
$$

In other words, for any point in time and space, the generalized distance between the approximated and the real value as seen by the sink is $\sqrt{(x - x_0)^2 + (\gamma kT)^2}$, and the corresponding *distortion per field point* of point x as seen by the sink is given by $2(1 - \sigma(|x - x_0|, kT))$.

In general, the statistics of the correlated data field might not be known, but they can be measured on-line during the network deployment period (for instance, if the correlation is distance dependent, then nodes can make use of the distance information acquired from the neighbors for constructing routing tables for the additional tasks of estimating the correlation structure and fitting the measurement parameters to a valid correlation model[7]).

6.5.2.3 Total Distortion

We compute now the total distortion of the data estimated by the sink at a snapshot in time in MSE sense. Consider node k, which is placed k hops away from the sink. The position of node k is denoted as x_k and the data that node k measures at time t as $X(x_k, t)$. The data sent at time t to the sink about the region $[x_k - d/2, x_k + d/2]$ are $X(x_k, t)$, but since it is *delayed* with k clock ticks, this packet actually reaches the sink at time $t + kT$ (see Figure 6.21). In fact, at time t, the actual available data at the sink about node k are $X(x_k, t - kT)$. Thus, the corresponding distortion of reconstruction of the region covered by node k is

$$D_k(N) = 4 \int_{x=x_k}^{x_k + \frac{L}{2N}} [1 - \sigma(x - x_k, kT)]\, dx. \qquad (6.22)$$

For simplification, we can consider $x_k = 0$ as the axis origin for each node k, and then (6.22) becomes

$$D_k(N) = 4 \int_{x=0}^{\frac{L}{2N}} [1 - \sigma(x, kT)]\, dx. \qquad (6.23)$$

The total distortion $D(N)$ is simply obtained by summing (6.23) over all nodes $k = 0 \dots N - 1$:

$$D(N) = \sum_{k=0}^{N-1} 4 \int_{x=0}^{\frac{L}{2N}} [1 - \sigma(x, kT)]\, dx. \qquad (6.24)$$

Further, if we insert the correlation model for a Gaussian spatio-temporal process (6.1), we can finally write the total distortion of reconstruction of the whole field by the sink as a function of N:

$$D(N) = \sum_{k=0}^{N-1} 4 \int_{x=0}^{\frac{L}{2N}} \left\{ 1 - \exp[-c(x^2 + \gamma^2 (kT)^2)^{\frac{\beta}{2}}] \right\} dx, \qquad (6.25)$$

where k counts the number of hops from a node to the sink, T is the time delay per hop, γ is the time scaling constant, and c is a constant quantifying the intensity of correlation of the field. The term which is integrated is the distortion incurred by approximating the field, between $[-\frac{L}{2N}, \frac{L}{2N}]$, around the node which is at k hops away from the sink, with the value of that node delayed k time steps.

The expression in (6.25) cannot be provided in a closed form. However, an experimental analysis will show that (6.25) has always a minimum as a function of N. Moreover, in Section 6.5.2, we use a strong correlation approximation to derive in a closed form the optimal value N for which (6.25) is minimized. In general, the optimal value of N is obtained by setting $\frac{\delta D(N)}{N} = 0$ and numerically solving for N and by rounding the solution to the closest integer.

Since spatially correlated Gaussian processes can only have certain structures for the correlation dependence on the distance,[7] we restrict our analysis to the models introduced in Section 6.3.2.

6.5.2.4   Optimum N is Finite

In this subsection we show that for the Gaussian correlation models introduced in Section 6.3.2, there is indeed a *finite* optimum N_0 that minimizes (6.25). Denote by

$$a_n = \sum_{k=0}^{n-1} \int_{x=0}^{\frac{L}{2n}} \sigma(x, kT)dx. \tag{6.26}$$

Note that by definition a_n is lower bounded by 0. Thus, from (6.24) it results a sufficient condition for the existence of a finite optimum N. Namely, the condition is that there exists N_0 such that for all $n > N_0$, a_n is a decreasing sequence.

(i) *Correlation model*: $\exp(-c[x^2 + \gamma^2(kT)^2])$
     In this case, we can rewrite a_n as

$$a_n = \int_{x=0}^{\frac{L}{2N}} e^{-cx^2} dx \cdot \sum_{k=0}^{n-1} \frac{1}{e^{\gamma^2(kT)^2}}. \tag{6.27}$$

But $\lim_{n\to\infty} a_n \to 0$, since the first term in the product converges to zero (the error function) and the second term can be easily shown to be upper bounded by a finite positive constant. Thus, for $\beta = 2$, the optimum N that minimizes (6.25) is finite.

(ii) *Correlation model*: $\exp[-c\sqrt{x^2 + \gamma^2(kT)^2}]$
     This case is difficult to analyze analytically, due to the function that is integrated. However, our simulations in Section 6.5.4 show that in this case too there is a finite optimal N_0.

6.5.2.5   Strong Correlation Approximation

We study the case when both L/N and the time scale γ are small. In other words, data are strongly correlated both spatially and temporally. In this case, we can make the approximation

$$1 - e^{-c[x^2 + \gamma^2(kT)^2]^{\frac{\beta}{2}}} \approx 1 - (1 - c[x^2 + \gamma^2(kT)^2]^{\frac{\beta}{2}})$$

$$= c[x^2 + \gamma^2(kT)^2]^{\frac{\beta}{2}},$$

which simplifies our analysis further.

(i) *Correlation model*: $\exp(-c[x^2 + \gamma^2(kT)^2])$

First, we can write

$$\int_{x=0}^{\frac{L}{2N}} c[x^2 + (\gamma kT)^2]dx = c\left[\frac{L^3}{24N^3} + \frac{L(\gamma kT)^2}{2N}\right]. \tag{6.28}$$

Further, from rewriting (6.25), we get

$$D(N) = c\left[\frac{L^3}{2}\cdot\frac{1}{N^2} + \frac{2L(\gamma T)^2}{3}\cdot N^2 - L(\gamma T)^2\cdot N + \frac{1}{3}L(\gamma T)^2\right].$$

Now we take the partial derivative of $D(N)$ with respect to N and make it equal to zero. Then, N_0 is a solution of the equation

$$4N^4 - 3N^3 - \alpha = 0, \tag{6.29}$$

where $\alpha = \frac{L^2}{\gamma^2 T^2}$. For $\alpha > 0$, (6.29) has a single positive solution N_0.

This gives a good indication of the optimal value of N_0 as a function of $\frac{L^2}{\gamma^2 T^2}$; intuitively, N_0 increases with the decrease of the importance of delay in the distortion function given by the time scale γ.

Note that, as expected, in the approximation of very strong correlation, the optimal N does not depend on the value of c (which models the strength of correlation). The optimal value of N only depends on the ratio between the length of the field L and the time scaling parameter γT, which models the relative importance of delay in the distortion function as compared to spatial distortion.

(ii) *Correlation model*: $\exp[-c\sqrt{x^2 + \gamma^2(kT)^2}]$

In this case, we compute

$$\int_{x=0}^{\frac{L}{2N}} c[x^2 + (\gamma kT)^2]^{\frac{1}{2}}dx = -\frac{c}{8N}\left[-L\sqrt{\frac{L^2 + 4(\gamma Tk)^2 N^2}{N^2}}\right.$$

$$+ 2(\gamma Tk)^2 N \ln(\gamma Tk)^2$$

$$+ 4(\gamma Tk)^2 N \ln 2$$

$$\left. - 4(\gamma Tk)^2 N \ln\frac{L + \sqrt{\frac{L^2 + 4(\gamma Tk)^2 N^2}{N^2}}N}{N}\right]. \tag{6.30}$$

When N is large, the second, third, and fourth terms in the summation in the paranthesis cancel out each other, and it can be easily shown that (6.30) simplifies to

$$\int_{x=0}^{\frac{L}{2N}} c[x^2 + (\gamma kT)^2]^{\frac{1}{2}} dx \approx \frac{cL\gamma Tk}{2N}. \tag{6.31}$$

By summing (6.31) over k, we can see that the resulting sum is a strictly increasing sequence in N. This only happens for N large enough to guarantee the strong correlation approximation; however, this is enough to show that the optimum N_0 has to be finite.

6.5.3   Two-Dimensional Networks

6.5.3.1   Total Distortion
The case of a two-dimensional grid network (see Figure 6.22) is studied similarly to the one-dimensional model. Consider a square area $L \times L$, on which N^2 nodes are uniformly placed on a square grid. The network is divided into Voronoi cells centered in the sensor nodes. We count the number of hops from each node to the sink on the most energy

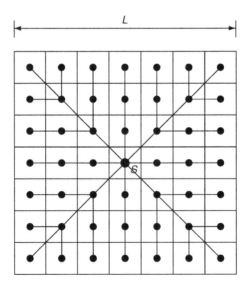

Figure 6.22 A two-dimensional square grid. The Voronoi cell partition is drawn is dashed lines, and the SPT is drawn in bold solid lines.

efficient transmission structure for gathering uncorrelated data, which is the SPT. Note that, in general, since data at nodes are spatially correlated, the SPT is *not* the most energy efficient transmission structure if in-network fusion by coding with side information is performed at nodes. Moreover, finding the optimal transmission structure for such scenarios is NP-hard.[9,12] Thus, in our analysis, we make the assumption that, due to limited resources, relay sensor nodes do not perform in-network fusion, namely, they do not use as side information data from other nodes that use them as relay. In short, data are relayed without being processed.

In order to simplify the analysis, we consider a slightly modified setting for the two-dimensional square grid as compared to the one-dimensional model (see Figure 6.22). The modification from the one-dimensional study is that, for the two-dimensional model, we assume that the sink gathers with no delay data in its corresponding Voronoi cell (in other words, the sink itself is considered as a regular sensor).

We plot in Figure 6.22 the energy efficient paths from the nodes to the sink. Note that for every $k = 0 \ldots N-1$, there are $8k$ cells situated at k hops away from the sink. Therefore, analogously to the one-dimensional case, we can write the total distortion for the two-dimensional case as

$$D(N) = \sum_{k=0}^{N-1} 4 \cdot 8k \int_{x=0}^{\frac{L}{2N}} \int_{y=0}^{\frac{L}{2N}} \left\{ 1 - \exp(-c[x^2 + y^2 + \gamma^2(kT)^2]^{\frac{\beta}{2}}) \right\} dy dx, \quad (6.32)$$

where N is now the number of hops from the sink to the extremity of the square network.

6.5.3.2 Strong Correlation Approximation

In this section we use an approximation similar to the one in Section 6.5.2, namely,

$$1 - \exp(-c[x^2 + y^2 + \gamma^2(kT)^2]^{\frac{\beta}{2}}) \approx c[x^2 + y^2 + \gamma^2(kT)^2]^{\frac{\beta}{2}}. \quad (6.33)$$

For the sake of simplicity, we analyze only the case $\beta = 2$, since the resulting optimization is easier. After some straightforward manipulations including taking the partial derivative of the resulting $D(N)$ with respect to N, we obtain that the optimal N_0 is a solution of the equation:

$$N^5 - N^4 - \frac{\alpha}{3}N + \frac{\alpha}{2} = 0 \quad (6.34)$$

where $\alpha = \frac{L^2}{\gamma^2 T^2}$. A full analysis of the behavior of this polynomial is outside the scope of this chapter. However, through numerical experiments, we are able to provide a set of insights:

- For $0 < \alpha < 83.9$, this equation has no real positive solution. It is strictly increasing, and thus, its optimum is attained at $N_0 = 1$, which means that the distortion caused by delay becomes so important that the optimal solution is to not place any sensor and let the sink estimate the whole field!

- For $\alpha \geq 83.9$, the equation has two positive real solutions, one ($N_1 \in (1, 2)$) corresponding to a maximum of the function $D(N)$, and the other $N_2 > 2$ to a minimum of the function. For $N > N_2$, $D(N)$ is strictly increasing. Thus, the optimum solution is either in $N_0 = 1$, or in N_2, both being finite integers.

6.5.4 Numerical Simulations

In this section, we do not use the approximation of strong correlation, but rather use the rough total distortion formulae given by equations (6.25) and (6.32).

We use Maple to plot in Figure 6.23(a) the distortion $D(N)$ for the one-dimensional case, as expressed in equation (6.25), as a function of

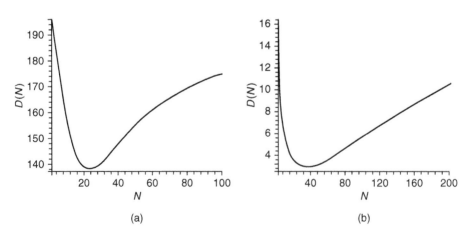

(a)   (b)

Figure 6.23 Total estimation distortion of the field at the sink $D(N)$ as a function of the number of nodes N for a one-dimensional network: (a) $\frac{\beta}{2} = 1$, and (b) $\frac{\beta}{2} = 0.5$.

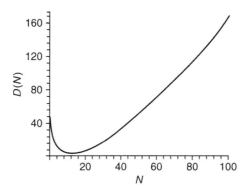

Figure 6.24   Total estimation distortion of the field at the sink $D(N)$ as a function of the number of nodes N for a two-dimensional network.

N, for typical values of the constants involved: $c = 0.5$ (reasonable correlation decay), $\gamma T = 0.1$ (the constant scaling the time axis), $L = 100$, and $\beta = 2$. In Figure 6.23(b) we illustrate with a similar plot the case when $\beta = 1$, with $c = 0.05$, $\gamma T = 0.05$, and $L = 100$. We observe that, in general, there is an optimal N that depends on the few constants involved in our model: c, γ, β, L.

Finally, in Figure 6.24, we plot the distortion $D(N)$ for the two-dimensional case, as expressed in equation (6.32), as a function of N, for typical values of the constants involved: $c = 0.05$, $\gamma T = 0.05$, $L = 10$, and $\beta = 2$. We observe that, again, there is an optimal N minimizing the total distortion.

6.5.5   Remarks

In this section, we studied data gathering of spatio-correlated processes from a correlated data field to a sink. The task was to perform at the sink the reconstruction of the data measured at all the points in the field, with maximal accuracy, when the only available information was the data at the sensor nodes. We defined a single measure of accuracy that combined the distortions due to the spatial approximation and to the delay in the network. We showed that, in general, there is a finite optimal density of sampling the field. Future work includes the analysis of more complex interpolation strategies and the study of a wider class of random processes.

6.6   Conclusion

We studied the interaction between data representation at nodes, rate allocation, and routing and node placement for gathering of correlated data in sensor networks. The results of our work show that a joint consideration of these issues provides important improvements in the overall data gathering energy efficiency and accuracy of representation.

We first analyzed energy efficient data gathering of random spatially correlated processes with lossless and lossy coding. We successively found the optimal transmission structure and rate-distortion allocations. Moreover, we considered the problem of energy efficient optimal node placement. Further, we considered data gathering of spatio-temporally correlated processes under delay constraints. We defined a distortion measure that includes the effects of both spatial approximation and delay. We showed that, in general, there is an optimal finite density of nodes that should be placed in the field for minimizing the total distortion of reconstruction at the sink.

In all the scenarios, the non-trivial interaction between the structure of the data (or underlying signal) and the transport mechanism to a central sink has been highlighted. In our view, this is a central challenge in the design and operation of sensor networks. No simple "separation" can be used, and substantial gains are obtained by a joint analysis and design.

References

1. A. Aaron and B. Girod, Compression with Side Information Using Turbo Codes, in *Proc. DCC*, 2002. Snowbird, UT.
2. J. Barros, C. Peraki, and S. D. Servetto, Efficient Network Architectures for Sensor Reachback, in *Proc. of the IEEE International Zurich Seminar on Communications*, 2004. Zurich, Switzerland.
3. T. Berger, Multiterminal Source Coding, Lecture notes presented at the CISM summer school, 1977.
4. T. Berger and J. D. Gibson Lossy Source Coding, *IEEE Trans. Inf. Theory*, 1998. 44(6): 2693–2723.
5. D. Bertsekas, *Network Optimization: Continuous and Discrete Models*, Athena Scientific, Belmont, MA, 1998.
6. T. M. Cover and J. A. Thomas, *Elements of Information Theory*, John Wiley & Sons, New York, 1991.

7. N. Cressie, *Statistics for Spatial Data*, Wiley, New York, 1991.
8. R. Cristescu and B. Beferull-Lozano, Lossy Network Correlated Data Gathering with High-Resolution Coding, to appear IEEE Trans. on Information Theory, 2006.
9. R. Cristescu, B. Beferull-Lozano, and M. Vetterli, On Network Correlated Data Gathering, in *Proc. INFOCOM*, 2004. Hong Kong, PR China.
10. R. Cristescu, B. Beferull-Lozano, and M. Vetterli, *Networked Slepian-Wolf: Theory, Algorithms and Scaling Laws, IEEE Trans. on Information Theory*, 2003. 51(12): 4057–4073.
11. R. Cristescu, B. Beferull-Lozano, and M. Vetterli, Scaling Laws for Correlated Data Gathering, in *Proc. ISIT 2004*, Chicago, IL, June 2004.
12. R. Cristescu, B. Beferull-Lozano, M. Vetterli, and R. Wattenhofer, *Network: Correlated Data Gathering with Explicit Communication: NP-Completeness and Algorithms, IEEE/ACM Trans. on Networking*, in press, to appear April 2006.
13. R. Cristescu and M. Vetterli, On the Optimal Density for Real-Time Data Gathering of Spatio-Temporal Processes in Sensor Networks, in *Proc. IPSN*, 2005. Los Angeles, CA.
14. R. Cristescu and M. Vetterli, Power Efficient Gathering of Correlated Data: Optimization, NP-Completeness and Heuristics, in *Proc. MobiHoc 2003* (poster), Annapolis, MD, June 2003.
15. M. Enachescu, A. Goel, R. Govindan, and R. Motwani, Scale Free Aggregation in Sensor Networks, *Algosensors*, 2004. Turku, Findland.
16. D. Ganesan, R. Cristescu, and B. Beferull-Lozano, Power-Efficient Sensor Placement and Transmission Structure for Data Gathering under Distortion Constraints, *ACM Trans. on Sensor Networks*, to appear 2006.
17. D. Ganesan, S. Ratnasamy, H. Wang, and D. Estrin, Coping with Irregular Spatio-Temporal Sampling in Sensor Networks, in *Proc. HotNets-II*, 2003. Held at MIT, Cambridge, MA.
18. M. Gastpar, B. Rimoldi, and M. Vetterli, To Code, or Not to Code: Lossy Source-Channel Communication Revisited, *IEEE Trans. Inf. Theory*, 2003. 49(5): 1147–1158.
19. M. Gastpar and M. Vetterli, Power, Spatio-Temporal Bandwidth, and Distortion in Large Sensor Networks, *IEEE J. Selected Areas Commun.*, Vol. 23, No. 4, April 2005.
20. A. Goel and D. Estrin, Simultaneous Optimization for Concave Costs: Single Sink Aggregation or Single Source Buy-at-Bulk, *ACM-SIAM Symp. on Discrete Alg.*, 2003. Baltimore, MD.
21. W. Rabiner Heinzelman, A. Chandrakasan, and H. Balakrishnan, Energy-Efficient Communication Protocol for Wireless Microsensor Networks, in *Proc. HICSS*, 2000. Hawaii.
22. A. Kumar, P. Ishwar, and K. Ramchandran, On Distributed Sampling of Bandlimited and Non-Bandlimited Sensor Fields, in *Proc. ICASSP*, 2004. Montreal, Canada.

23. H. Larson and B. Shubert, *Probabilistic Models in Engineering Sciences*, John Wiley & Sons, New York, 1979.
24. S. Lindsey, C. S. Raghavendra, and K. Sivalingam, Data Gathering in Sensor Networks Using the Energy*Delay Metric, in *IEEE Transactions on Parallel and Distributive Systems*, special issue on Mobile Computing, pp. 924–935, April 2002.
25. P. S. Maybeck and S. Peter, Stochastic Models, Estimation, and Control, *Mathematics Science Engineering*, Vol. 141, 1979.
26. D. Marco, E. Duarte-Melo, M. Liu, and D. L. Neuhoff, On the Many-to-One Transport Capacity of a Dense Wireless Sensor Network and the Compressibility of Its Data, in *Proc. IPSN*, 2003. Palo Alto, CA.
27. J. Massey, Joint-Source Channel Coding, in *Communication Systems and Random Process Theory*, 1978. pp. 279–293.
28. S. Pattem, B. Krishnamachari, and R. Govindan, The Impact of Spatial Correlation on Routing with Compression in Wireless Sensor Networks, in *Proc. IPSN*, 2004. Berkeley, CA.
29. B. Porat, *Digital Processing of Random Signals*, Prentice Hall, New York, 1994.
30. G. J. Pottie and W. J. Kaiser, Wireless Integrated Sensor Networks, *Commun. ACM*, 2000.
31. S. Pradhan, Distributed Source Coding Using Syndromes (DISCUS), Ph.D. thesis, U.C. Berkeley, CA, 2001.
32. S. Pradhan and K. Ramchandran, Distributed Source Coding Using Syndromes: Design and Construction, in *Proc. DCC*, 1999. Snowbird, UT.
33. A. Scaglione and S.D. Servetto, On the Interdependence of Routing and Data Compression in Multi-Hop Sensor Networks, in *MobiCom*, 2002. Atlanta, GA.
34. C. E. Shannon, A Mathematical Theory of Communication, *Bell Sys. Tech. J.*, 1948.
35. D. Slepian and J. K. Wolf, Noiseless Coding of Correlated Information Sources, *IEEE Trans. Inf. Theory*, 1973.
36. S. Vembu, S. Verdu, and Y. Steinberg, The Source-Channel Separation Theorem Revisited, *IEEE Trans. Inf. Theory*, 1995.
37. R. Zamir and T. Berger, Multiterminal Source Coding with High Resolution, *IEEE Trans. Inf. Theory*, 1999.

7

Wireless Networks and the Expected Next Revolution in Information Technology

P. R. Kumar*

7.1 Introduction

Over the past few years the problem of how to address wireless networks in the aggregate has been addressed, with the goal of making some progress toward understanding what their capabilities are and how they are to be operated. The issue is that every wireless network can potentially be different from others in the number of nodes it has, where they are located, the traffic requirements imposed on it, etc. Yet, one wants to establish some properties that will help in the evaluation and design of wireless networks. So motivated, we study models where distance plays a somewhat more explicit role than is traditional. Besides modelling the networks, the introduction of distance also allows us to quantify the capabilities of such networks.

To operate wireless networks in a multi-hop fashion, protocols for medium access control, routing, power control, and transport are

* This material is based upon work partially supported by DARPA/AFOSR under Contract No. F49620-02-1-0325, AFOSR under Contract No. F49620-02-1-0217, USARO under Contract Nos. DAAD19-00-1-0466 and DAAD19-01010-465, NSF under Contract Nos. NSF ANI 02-21357 and CCR-0325716, and DARPA under Contact Nos. N00014-0-1-1-0576 and F33615-0-1-C-1905.

needed. Since bandwidth in a wireless network may be scarce in comparison to optical fiber there is interest in increasing the efficiency of protocols for these networks. Thus, there have been efforts in advancing a "next" generation protocol suite that utilizes the shared wireless medium in a more efficient way. This effort has also spurred an interest in cross-layer design.

In addition to their utilization as ad hoc networks or mesh networks, wireless networks are also of interest for their deployment as wireless sensor networks. Nodes possessing sensing capability (temperature, light, magnetic, acoustic, etc.), that can also compute, store, and wirelessly communicate, can be deployed over domains to perform tasks such as environmental monitoring. The analysis and design of these sensor networks bring fresh challenges due to the new model of computation that requires an integrated treatment of the tradeoffs between local computation and message passing.

Actuation will, one may well suspect, closely follow sensing, and this leads to the next frontier of networked control. Performing control over shared computation and communication networks can lead to large-scale deployment of control systems. A fundamental challenge is what might be the appropriate abstractions, and what is an appropriate architecture that facilitates design, development, and deployment of these systems and enables their proliferation.

7.2   The Wireless Medium

Over the past few years there has been increasing interest in what are now called *ad hoc networks*; in the past they were also referred to as *packet radio networks*. Essentially, these are networks formed by nodes with radios. The proposed mode of operation[1] is through multi-hop relaying, shown in Figure 7.1. Packets are relayed from node to node in short hops until they reach their destination. An advantage of such networks is that they require no prior wired infrastructure. A set of users with laptops equipped with wireless PCMCIA cards can spontaneously form a network at an airport or on a university campus. Also of interest are so-called "mesh networks" that can connect residents of a community. Another advantage is that potentially these networks can also be deployed in mobile environments that preclude the use of wires. In fact, "wires" which originally facilitated the current information technology revolution have, in some of the envisaged applications, become barriers to proliferation.

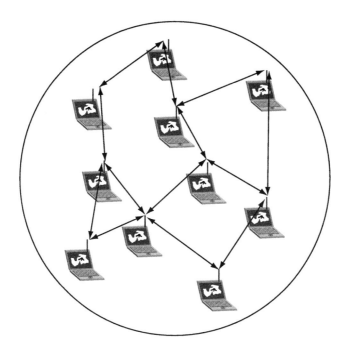

Figure 7.1   Ad hoc wireless networks.

The wireless medium is different from the wired medium in several respects. First, it is a *shared* medium. That is, users can interfere with each other. Users will therefore need to cooperate with each other to avoid interfering with each other's packets. Figure 7.2 shows what is called a "hidden" terminal scenario. Node A is sending a packet to node B. However, due to node C, a neighbor of node B that is also sending a packet to node D, there is a "packet collision," resulting in the non-reception of the packet from node A by node B. Indeed, addressing packet collisions has been a topic central to the study of wireless networks since the early days of the ALOHA network.[2] One solution to this is to use *carrier sensing*, a technique used on the Ethernet. Potential transmitters listen for the carrier to see if any other nodes are transmitting before they begin their transmission. However, this is not a perfect solution in wireless networks. Figure 7.3 shows a scenario called the "exposed" terminal problem. Node A wants to

Figure 7.2   A hidden terminal scenario.

Figure 7.3 The exposed terminal problem.

send a packet to node B. However, it hears node C's carrier and so refrains from transmitting. The "tragedy" is that the slot is wasted, since node C is remote from node B, and so its transmission does not destructively interfere with the reception by node B of the packet from node A. It should be clear from these examples that "optimally" sharing the wireless medium is not a simple task. This is the problem of *medium access control* to which we return in Section 7.3.3.

The shared nature of the wireless medium, resulting in nodes potentially getting reduced throughput due to the presence of other transmitting nodes, necessitates an understanding of *how much traffic* wireless networks can carry. This is the topic of Section 7.2.

A second sense in which wireless networks can be (but need not always be) different from wired networks is that often the data rates of wireless "links" are lower (often orders of magnitude lower) than optical pipes. Thus, users sharing a link do suffer a throughput reduction. This again necessitates efficient utilization of the wireless medium. Consider, for example, the *routing problem*. As shown in Figure 7.4, the traffic from node *s* to node *d* may find it more advantageous to follow a more circuitous path to avoid crisscrossing the path of packets traversing along an already established path. We will return to this issue in Section 7.3.2.

There is another manner in which the wireless medium differs from the wired medium. It can be much more "unreliable." For example, due to electromagnetic interference from noise sources such as microwave ovens (which actually co-occupy the 2.4 GHz band with IEEE 802.11b

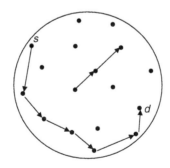

Figure 7.4 Flow avoiding routing.

cards), packets can be lost. Also, distinguishing features of the wireless medium are multi-path effects, Doppler shifts due to mobility, and fading. This can necessitate measures at lower layers to enhance reliability, such as link-level acknowledgments. This has an effect of the design of the medium access control layer.

The potential capability of creating *mobile* networks itself introduces some challenges. One is that of locating a destination node to which packets have to be delivered — the problem of *routing*. Much traditional work in routing has been focused on minimum hop routing, which can be solved through distributed Bellman-Ford type algorithms.[3] In a dynamic environment, this algorithm can lead to loops. Consequently, techniques using sequence numbers have been developed[4] to avoid such loops. Several routing algorithms have made it to the IETF Internet Draft status.[1] As shown in Figure 7.4, there can be advantages in routing that take account of either pre-existing routes or the current "load" on the network and adapts to it. Due to the paucity of wireless link data rates, there may even be a benefit to multi-path routing, as shown in Figure 7.5. Multiple routes from a source to a destination d can potentially be simultaneously utilized to provide greater end-to-end throughput. We examine the issue of combining both multi-path routing and load adaptive routing in Section 7.3.

In addition to serving as data networks, wireless networks can also be deployed as *wireless sensor networks*. Nodes can be attached to acoustic, temperature, magnetic, light, or other sensors. They can also be endowed with computation capability. When they are also wirelessly interconnected, they can be used to form wireless sensor networks. Indeed, nodes comprising hardware for such sensor networks called "motes" have been developed,[5] along with an appropriate operating system.[6]

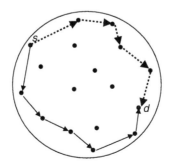

Figure 7.5 Multi-path routing.

Such sensor networks can be used for monitoring the environment[7] or object localization and tracking.[8] These networks are frequently application specific; that is, the entire network is deployed to provide a specific functionality. Because of this, messages received and/or sensor readings taken at motes may trigger some local computation which then triggers some packet generation and transmission. This makes sensor networks different from data networks. In data networks, the "payload" of a packet is sacrosanct, and intermediate relay nodes do not alter the payload. In contrast, in sensor networks, the goal is not merely delivering packets from one node to another, but may also be to detect/inform a collector or sink node when some event of interest happens. For example, the sink node may only be interested in being notified when the maximum sensor reading in the network exceeds some threshold. Thus, the sink need not get the readings of all the measurements of all the sensors in the network, but only, say, the maximum. Given this limited interest in the data, intermediate nodes may therefore fuse data, or even drop it altogether if it is not interesting, rather than faithfully relaying it. This critical difference leads nodes to make computations based on incoming packets as well as their own readings and to make judicious decisions on what to themselves transmit.

The result is that the entire network can be viewed as a distributed computer, where nodes are interconnected by the wireless medium.

Another sometimes distinguishing feature of sensor networks is that they may often operate under severe resource constraints. It may be required that a sensor network be deployed in an unattended remote habitat for several months. Each node may only have a pair of AA batteries. Thus, the network itself may be very energy constrained. Also, computing hardware at the nodes may have very limited memory and computational capabilities.

The net result is that we face the problem of studying quite different models of computation than is traditionally the case. This requires a theory of *in-network processing* that informs us how best to use the computation and communication capabilities for the specific task that the sensor network has been developed. We visit this issue in Section 7.4.

Sensing the environment leads inexorably, we believe, to actuation; that is, one wants to exploit the ability to sense and measure the environment in order to act on it and alter it. Since sensing and actuation together comprise what is called "control," we believe that control over networks could be a possible next phase of the information technology revolution. It will lead to the convergence (or reconvergence, see Mindell[9]) of control with communication and computation.

While we have noted that sometimes sensor networks may also be resource starved in terms of energy, computation, and communication capabilities, there are also important environments where all three may be plentiful. Such "course granularity" deployments may be an essential component of future networked control systems, be they transportation systems, smart energy grids, or air transportation management.

In such complex deployments, a crucial resource is the designer's time or, more generally, the design-develop-deploy cycle-time. Lengthy development periods can lead to huge development costs which may make deployment prohibitively expensive, stifling the proliferation of networked control.

To enhance proliferation, we believe that a fundamental challenge is to determine what the appropriate *abstractions* are and what the appropriate *architecture* of such systems is. Indeed, we argue in Section 7.5 that architecture and abstractions are important for proliferation of technology.

Our goal is to support such abstraction through middleware. This is analogous to supporting the abstraction of a "graph" provided by the network layer in the OSI stack (see Figure 7.6), which is supported by

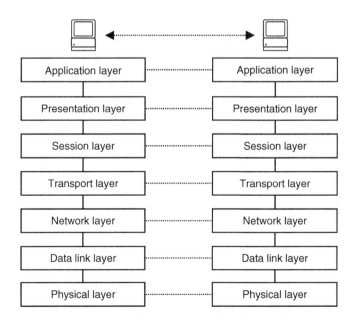

Figure 7.6   The layers of the OSI stack.

routing protocols. Similarly, we believe that by providing appropriate middleware for networked control, we can simplify the development of networked control. We will examine this issue in Section 7.5.

7.3 The Capacity and Architecture of Wireless Networks

It is convenient to begin with a very simple model of a wireless network that can be embellished with increasingly more accurate physical descriptions.

Consider a domain of area A square meters as shown in Figure 7.7. Within A are located n nodes in arbitrary locations. As noted in Section 7.3, a distinguishing characteristic of the wireless medium is that it is a shared medium. Thus, we need to begin by defining when such sharing is successful. For this purpose we consider a simple (even simplistic) model. Suppose, as in Figure 7.8 that node T wishes to send

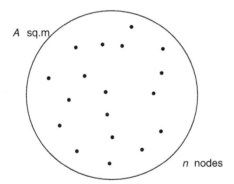

Figure 7.7 A domain of area A with n nodes.

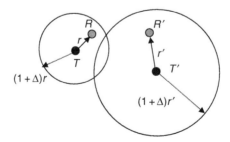

Figure 7.8 The protocol model.

a packet to node R at a distance r. We will simply suppose that node T scales its power appropriately and transmits the packet. The critical assumption we make is that such a transmission of range r creates an interference footprint of radius $(1 + \Delta)r$ centered around the transmitter. Here, $\Delta > 0$ is some positive parameter. Within this "interference footprint" of the transmission, we suppose that no receiver located in it can potentially pick up any other transmission. Suppose as in Figure 7.8 that there is another concurrent transmission by node T' to node R' at a distance r' away. Then that transmission, in turn, creates an interference footprint centered around T' of radius $(1 + \Delta)r'$. Now in order for the transmission from T to R to be successful, we need the distance from T' to R be larger than or equal to $(1 + \Delta)r'$. Similarly, for the transmission from T' to R' to be successful, we need the distance from T to R' be larger than or equal to $(1 + \Delta)r$.

This type of model for when concurrent transmissions are successful has been called the *Protocol Model* in Gupta and Kumar.[10] The reason is that some protocols may enforce or lead to such a constraint. Consider, for example, a cellular system where a frequency used in one cell may not be reused in an adjoining cell. For hexagonal cells, as illustrated in Figure 7.9, this is equivalent to a choice of (nearly) $\Delta = 2$.[**]

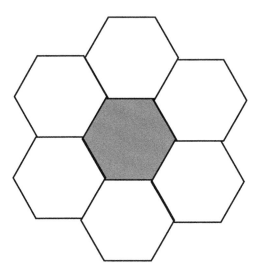

Figure 7.9   Frequency reuse prohibited in adjoining cells.

[**] We are implicitly assuming isotropic transmitting antennas in the above; however, similar results can also be obtained for directioal antennas (see Agarwal and Kumar[11]).

More formally, if $X_1, X_2, \ldots, X_n$ are the locations of n nodes in the domain, and if $X_i \to X_j$ and $X_k \to X_\ell$ are concurrent transmissions, then we require that

$$|X_i - X_\ell| \geq (1 + \Delta)|X_i - X_j|, \text{ and} \tag{7.1}$$

$$|X_k - X_j| \geq (1 + \Delta)|X_k - X_\ell|, \tag{7.2}$$

in order for both transmissions to be successful.

Concerning the quantity of information that can be transmitted, we simply suppose that each successful transmission takes place at a data rate of W bits per second.

What we have provided is a simple model which is certainly not an information-theoretic model. Rather, it is more of a model of how technology and protocols may be utilized.

Nevertheless, in spite of its simplicity, one can draw several useful conclusions. What makes these conclusions important is that they continue to hold under more physically meaningful models.

Note that the triangle inequality for distances between points in the plane says that

$$|X_i - X_\ell| \leq |X_i - X_j| + |X_j - X_\ell|. \tag{7.3}$$

Similarly,

$$|X_k - X_j| \leq |X_k - X_\ell| + |X_\ell - X_j|. \tag{7.4}$$

Thus, from (7.3) and (7.1), we obtain

$$|X_j - X_\ell| \geq |X_i - X_\ell| - |X_i - X_j| \qquad \text{(from (7.3))}$$

$$\geq (1 + \Delta)|X_i - X_j| - |X_i - X_j| \qquad \text{(from (7.1))}$$

$$= \Delta|X_i - X_j|. \tag{7.5}$$

Similarly, from (7.4) and (7.2), we obtain

$$|X_\ell - X_j| \geq |X_k - X_j| - |X_k - X_\ell| \qquad \text{(from (7.4))}$$

$$\geq (1 + \Delta)|X_k - X_\ell| - |X_k - X_\ell| \qquad \text{(from (7.2))}$$

$$= \Delta|X_k - X_\ell|. \tag{7.6}$$

Adding (7.5) and (7.6), we obtain

$$|X_j - X_\ell| \geq \frac{\Delta|X_i - X_j|}{2} + \frac{\Delta|X_k - X_\ell|}{2}. \tag{7.7}$$

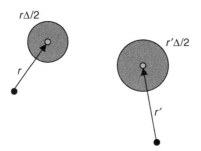

Figure 7.10 Virtual space used by transmissions.

This can be visualized as in Figure 7.10. If we draw a disk of radius $\frac{\Delta}{2}|X_i - X_j|$ around the receiver X_j, and a disk of radius $\frac{\Delta}{2}|X_k - X_\ell|$ around the receiver node X_ℓ, then these disks are essentially disjoint (the area of their intersection is zero). This has a very nice interpretation. We can imagine that each transmission of range r uses up a virtual space consisting of a disk of radius $\frac{r\Delta}{2}$ centered around its receiver. To elaborate a bit further, what it implies is that area is a valuable resource in wireless networks, in addition to spectrum. Indeed, in trying to economize on both of them, we obtain the well-known strategy of spatial reuse of spectrum, illustrated in Figure 7.9. The frequency spectrum used in a certain cell is reused in another distant cell.

Now recall that we are assuming a domain of area A square meters. Let $\{X_i : i \in T \subseteq \{1, 2, \ldots, n\}\}$ be the set of concurrent transmissions, with the range of the transmission from X_i being r_i. Then each such transmission uses an area of at $\frac{\pi r_i^2 \Delta^2}{16}$, as shown in Figure 7.11. (The area of the disk around each receiver is only $\frac{\pi r_i^2 \Delta^2}{4}$, and the extra factor of 4 in the denominator comes in because not all of this disk need lie in the domain.) Thus, we have the fundamental *spatial constraint*

$$\frac{\pi \Delta^2}{16} \sum_{i \in T} r_i^2 \leq A.$$

This can be written as

$$\frac{1}{T} \sum_{i \in T} r_i^2 \leq \frac{16A}{T\pi\Delta^2}, \tag{7.8}$$

where we have abused notation and used T to denote the cardinality of the set T.

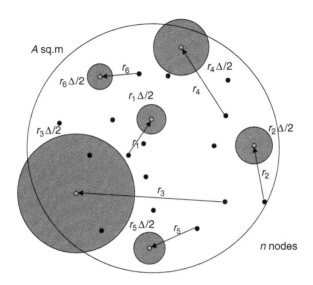

Figure 7.11   The space used by all the concurrent transmissions.

Now recall that due to the convexity of the function r_i^2, one has

$$\left(\frac{1}{T}\sum_{i\in T} r_i\right)^2 \le \frac{1}{T}\sum_{i\in T} r_i^2. \tag{7.9}$$

So we have, from (7.8) and (7.9),

$$\frac{1}{T}\sum_{i\in T} r_i \le \sqrt{\frac{16A}{T\pi\Delta^2}}.$$

This leads to

$$\sum_{i\in T} r_i \le \sqrt{\frac{16AT}{\pi\Delta^2}}.$$

Now the number of transmissions cannot exceed $\frac{n}{2}$, i.e.,

$$T \le \frac{n}{2},$$

since, at most, half the nodes can be transmitters, while the rest are receivers. Hence, we obtain

$$\sum_{i\in T} r_i \le c_1\sqrt{An}, \tag{7.10}$$

where

$$c_1 := \sqrt{\frac{8}{\pi\Delta^2}}.$$

We interpret $\sum_{i\in T} r_i$ as follows. When one bit has been transferred a distance of r_i meters, let us say that r_i *bit-meters* have been pumped.

Recalling that every node can transmit at W bits per second when it is successfully transmitting, we see that the number of bit-meters per second being pumped by the entire network is $W\sum_{i\in T} r_i$, and this is bounded through (7.10) by

$$W\sum_{i\in T} r_i \leq c_1 W\sqrt{An}.$$

Since this is true no matter which feasible set of successful concurrent transmissions T has been chosen, we obtain the following result.[10]

Theorem 7.1   *The number of bit-meters per second that the wireless network can pump is upper bounded by $c_1 W\sqrt{An}$ bit meters per second, where*

$$c_1 := \sqrt{\frac{8}{\pi\Delta^2}}.$$

This quantity, bit-meters per second, that a network can pump at best is called the *transport-capacity of the network*.

Theorem 7.1 provides an upper bound on the transport capacity. It is easy to construct scenarios of node locations where a transport capacity of this asymptotic order in n can also be achieved. Figure 7.12 shows a configuration where a capacity of order $\Theta(W\sqrt{An})$ is, in fact, achieved.[†]

The pre-constant c_1 in the upper bound, and a corresponding pre-constant in the lower bound on maximal transport-capacity, can be improved. The bounds can be sharpened to lie within a factor $\sqrt{8}$ of each other (see Agarwal and Kumar[11]).

The above bound has been shown to hold under more physically meaningful models than the Protocol Model. One can consider Signal-to-Interference Ratio (SINR)-based models, and the $\Theta(\sqrt{n})$ bound can

---

[†] Here and in the sequel, we use Knuth's solution, popular in computer science. If $\sup_n |\frac{f(n)}{g(n)}| < +\infty$, we say that $f = O(f)$. If $f = O(g)$ as well as $g = 0(f)$, we say that $f = \Theta(g)$.

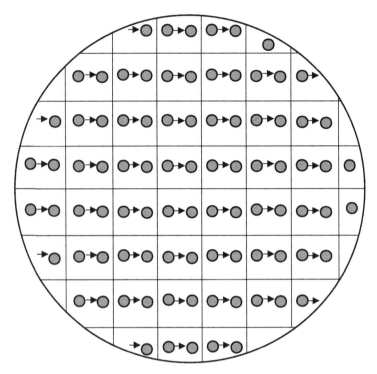

Figure 7.12 A regular arrangement of transmitters and receivers that pumps $\Theta(W\sqrt{An})$ bit-meters per second.

still be shown to hold (see Agarwal and Kumar[12]). Even the SINR-based model can be challenged, since one could argue that by making use of multi-user receivers, reception may be possible through successive decoding,[13] even at low SINRs. A convincing bound can only come from information theory. This has been pursued in Xie and Kumar,[14] where an analogous bound is shown. The case of fading has also been studied, and similar results continue to hold (see Xue, Xie, and Kumar[15]).

7.4 Protocols for Ad Hoc Networks

In order to realize the multi-hop mode of operation, several protocols are needed. A protocol that plays a critical role is *power control*. As we have seen in Section 7.2, the choice of appropriately low power range for transmissions is important to realize the capacity of wireless networks in the Protocol Model. The question then arises as to how this is to be done. How should nodes choose the powers of their packet transmissions? This is the *power control problem*.

Next, due to the choice of several short hops rather than one long hop (if that were, in fact, feasible), one has to find the sequence of nodes through which the packet is to be relayed. This is the *routing problem*.

Another protocol is *medium access control*. Since wireless is a shared medium, and packets can collide, at least under simple models of operation, nodes have to cooperate in preventing collisions and allowing data transfer to take place.

7.4.1 Power Control

We begin with the power control problem. The first issue that confronts us is to decide the protocol layer in which the power control problem is to be solved; the OSI stack is shown in Figure 7.6. There are several advantages to layering. The hierarchical layering allows segregation of development tasks. It ensures that individual efforts at the transport layer, network layer, physical layer, etc. can be pursued separately; yet the entire stack will function when they are put together. It thus allows plug and play functionality. It also makes maintenance and evolution easier. By lending longevity to the basic architecture, it ensures that proliferation can take place over the long haul by providing design stability.

Thus, it is important to grapple with the issue of where power control ought to be located vis-à-vis the protocol stack. It is at this point that problems arise. The choice of power level simultaneously affects signal quality, range, and congestion. Thus, one could argue based on its affecting signal quality that it is a physical layer issue. Indeed, this is effectively the approach pursued in cellular systems. However, it could also be argued that it is a network layer issue because the choice of power level affects range which, in turn, determines routes. On the other hand, it could also be argued that since excessive power levels by a transmitter cause "congestion" to a receiver which is not the intended recipient of its transmission, power control ought to be a transport layer issue. So where should power control be addressed?

Our contention made in Narayanaswamy et al.[16] is that power control in wireless ad hoc networks ought to be treated as a network layer issue. A brief summary of the reason is provided in Figure 7.13. From the analysis of capacity in Section 7.2, it follows that taking several

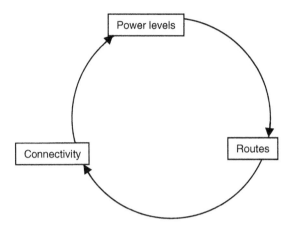

Figure 7.13   Mutual dependence of power levels, routes, and connectivity.

short hops is good. However, if the range of transmissions is too small, then the network can get disconnected. The reason is that, given the locations of the nodes, the choice of a value for the transmission range causes only those links to be formed whose length is no more than the transmission range, and the resultant collection of links may not yield a graph in which all nodes are connected. Thus, nodes should choose their power levels high enough to ensure connectivity. However, it is not possible to determine connectivity unless one has routes between nodes. Yet, one cannot choose routes without choosing power levels. Hence, as depicted in Figure 7.13, there is a mutual dependence between power control, connectivity, and routing.

It follows from the above that power control cannot be solved before the network layer, at least if one wants to perform a global rather than link-level optimization as, for example, necessitated by capacity considerations, since connectivity is determined only at the network layer. On the other hand, routing needs to be solved at the network layer, and it requires power control. This argues for a joint solution for power control and routing at the network layer.

Such a solution has been proposed and implemented in Narayanaswamy et al.[16] It pursues the approach of choosing what is the lowest *common* power level to be chosen by all nodes so that connectivity is maintained. The protocol is called COMPOW.

This protocol allows a clean modular implementation in conjunction with table-driven routing protocols. PCMCIA cards currently available off-the-shelf, for example, offer a choice of a handful of power levels ranging from 1 to 100 mW. Consider now running several

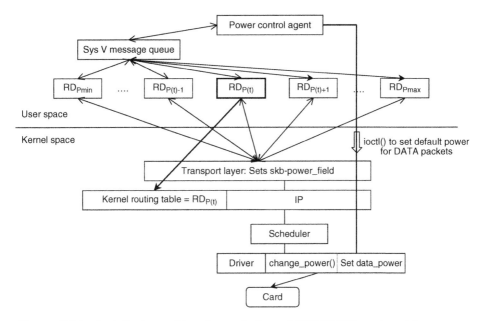

Figure 7.14   Architecture of implementation of COMPOW protocol for power control.

routing daemons in user space, one at each power level, as shown in Figure 7.14. From the examination of the routing tables for each power level, one can determine how many nodes are in the connected component. From this, it can be determined what is the lowest power level at which the number of nodes in the connected component is the same as at the highest possible power level. The routing table for this power level is simply copied over to the kernel routing table. The net result is a clean implementation at the network layer for what initially seemed like a cross-layer problem.

This protocol can be extended to also perform automatic clustering. The resulting CLUSTERPOW protocol is described in Kawadia and Kumar.[17] A more general treatment of power control can be found in Kawadia and Kumar.[18]

7.4.2   Routing

Over the past few years several routing protocols have advanced to the IETF Internet Draft stage.[1] These include both proactive protocols, that actively maintain routing tables, as well as reactive protocols that search for a route only when needed. Routing protocols can also be classified as distance-vector based, which function in a Bellman-Ford type of manner,[3] or link-state based,[19] where nodes attempt to develop more global knowledge.

What we will investigate in this section is load adaptive routing. As shown in Figure 7.5, nodes can benefit from using multiple routes along which to simultaneously route packets so as to get higher throughput. Also, as shown in Figure 7.4, routes can be made adaptive to other flows in the network and the load on the network, in order to choose good routes.

One approach proposed in Gupta and Kumar[20] is to adapt routes to the delay experienced by packets. The goal is to distribute the flow among several paths, as illustrated in Figure 7.15, so that the end-to-end delay experienced along each of the utilized paths is the same, and at the same time no more than the potential end-to-end delay over any unutilized path. Such a solution has been studied in the transportation literature and is known as a Wardrop equilibrium.[21]

The question is how to achieve this. The basic idea is to reduce flows over routes with delay in excess of the average delay in favor of flows with delay lower than the average delay. This, in turn, requires a delay estimation algorithm. The net result is a two-time scale adaptation

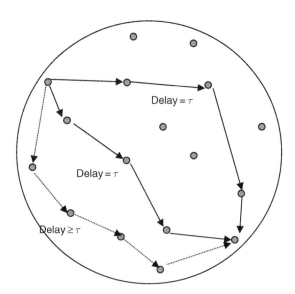

Figure 7.15 Wardrop routing.

scheme. Such an algorithm has been rigorously analyzed in Borkar and Kumar.[22]

The implementation of a multi-path delay adaptive algorithm poses several challenges. These have been studied in Raghunathan and Kumar.[23] First, it is desirable to improve the transient behavior of the algorithm prior to convergence. In particular, even in the transient phase one would like to avoid loopy routes for packets. Such a scheme is developed in Raghunathan and Kumar[24] based on the usage of even and odd fields. At even steps, packets are required to advance one hop closer to their destination, while at odd fields, they should move no further away in terms of hop count. Such a scheme is guaranteed to be loop free.

Additional issues also need to be resolved. An example is how to avoid the need for transport layer acknowledgment packets to carry delay feedback information needed for a network layer protocol. Some transport layer protocols, e.g., UDP, do not even have such acknowledgment (ACK) packets. These issues are described in Raghunathan and Kumar,[23] who provide implementation, simulation, and experimental testing results.

7.4.3 Medium Access Control

The last protocol we will touch upon is medium access control (MAC). The handshake portion of the now standard IEEE 802.11 protocol is illustrated in Figure 7.16. It consists of four phases. Initially, a transmitter announces its need to send a packet by sending a Request-to-Send (RTS) packet. The intended receiver then replies back with a Clear-to-Send (CTS) packet. The transmitter then replies with a DATA packet, following which the receiver sends an ACK. Nodes hearing an RTS or a CTS packet are required to refrain from transmitting for the duration, which ensures that collisions do not happen in either direction.

Additional mechanisms used include Carrier Sensing with a range larger than the decoding range for packets, as well as back-off counters that stagger and delay transmissions when conflicts do occur.

The question that we address is whether we can develop a MAC protocol that does not require such a four-phase handshake for each and every DATA packet on each and every hop, and one which does not require silencing of both the transmitter's and the receiver's

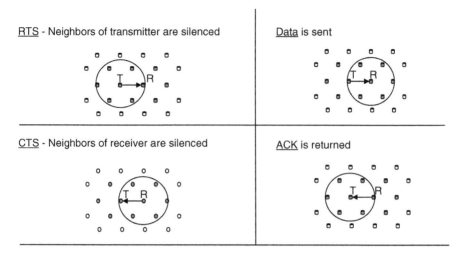

RTS - Neighbors of transmitter are silenced

Data is sent

CTS - Neighbors of receiver are silenced

ACK is returned

Figure 7.16 The four-phase handshake of IEEE 802.11.

neighborhoods. Finally, it may be desirable to replace the violent back-off mechanism with perhaps a gentler control.

Such a scheme has been examined in Rozovsky and Kumar.[25] It is a slotted protocol that uses random schedules for nodes involving either "listen" or "transmit-if-data" slots. Such schedules are developed through a pseudo-random number generator. In order for two nodes to know each other's schedules, they need only exchange the seeds of their random number generators. This results in the SEEDEX protocol. As shown in Figure 7.17, through a fan-in and fan-out procedure carried out at infrequent intervals, nodes determine the seeds of their

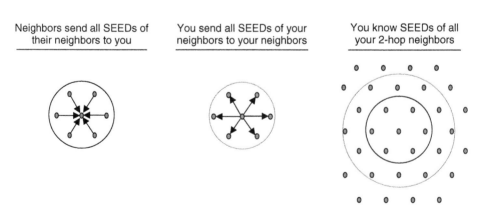

Neighbors send all SEEDs of their neighbors to you

You send all SEEDs of your neighbors to your neighbors

You know SEEDs of all your 2-hop neighbors

Figure 7.17 Fan-in and fan-out to determine seeds of two-hop neighbors.

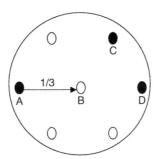

Figure 7.18 Probabilistic transmission based on two-hop neighbors' states.

two-hop neighbors. This is precisely the information that nodes need, since it is within this two-hop neighborhood that collisions originate (at least under some protocol models).

Once nodes know the seeds of their two-hop neighbors, they can take advantage of the knowledge of each other's schedules to reduce collisions. Consider the scenario shown in Figure 7.18, where node A wishes to send a packet to node B that, however, also has two other neighbors C and D in a transmit-if-data state. If all three nodes transmit, then there will be a collision. To alleviate this, node A randomly sends the packet with probability 1/3 in that slot to B. The choice of the value of 1/3 by node A is motivated by the fact that there are three contending neighbors of its receiver B, and is intended to optimize[‡] with respect to the number of neighbors likely to transmit.

This scheme can be modified in various ways. An additional multiplicative constant α can be incorporated so that the transmission probability is $\alpha/3$, where α is modulated according to the amount of load experienced at a node. Another possibility is to use SEEDEX only to make reservations for longer DATA packets.

The SEEDEX protocol, as well as the issues involved in implementing a slotted protocol on off-the-shelf cards, is described in Rozovsky and Kumar.[25]

[‡] However, strictly speaking, the choice of 1/3 may not be an optimal choice for node A, since nodes C and D may be desiring to transmit to nodes other than B and so may themselves be choosing probabilities differing from 1/3. In fact, this hints at some of the complexity of conducting performance evaluation of MAC protocols in wireless networks.

7.5   In-Network Processing for Function Computation in Sensor Networks

We now turn to sensor networks. In many such networks, there may be a *collector* or *sink* node, and the goal of the sink node is to determine some appropriate function of the sensor measurements taken at the other nodes.

For example, if $x_1, x_2, \ldots, x_n$ are the temperature measurements taken at n sensor nodes, then the sink node may want to obtain information on the *average* temperature $\frac{x_1 + x_2 + \cdots x_n}{n}$. Alternately, in an alarm network, the sink node may want to know the *maximum* temperature $\text{Max}_{1 \le i \le n} x_i$.

More generally, there may be a function $f(x_1, x_2, \ldots, x_n)$ that the sink node wants to compute. For many statistical functions such as mean, mode, median, max, standard deviation, histogram, etc., the function f is symmetric, i.e., invariant with respect to permutation of its arguments.

To address these issues one needs a model of the network. Consider a domain consisting of a disk of unit area with n nodes independently and uniformly distributed in it. Let $r(n)$ denote the transmission range of a node. We will suppose that all nodes choose the same value for their range. Consider the network formed by connecting any pair of nodes by a link when the distance between them is no more than $r(n)$. It is known that the resulting network is connected[26] with probability approaching one as $n \to +\infty$ if and only if

$$r(n) = \sqrt{\frac{\log n + k(n)}{\pi n}},$$

for some sequence $k(n)$ with $\lim_{n \to \infty} k(n) = +\infty$. Let us simplify matters by supposing that a range of $c\sqrt{\frac{\log n}{\pi n}}$ is used where $c > 1$. We will call this a random network. One node in this network is designated as the sink node. We will assume that nodes transfer packets to each other over such a network governed by the Protocol Model.

The values of the measurements are taken to lie in a finite alphabet. Assuming that nodes can compute and communicate, how should the entire network be operated to compute the function f? This problem has been studied in Giridhar and Kumar.[28]

The performance metric we address is the *computational throughput*, which is the rate at which the function f can be computed by the sink node. To maximize this, we allow block computation, as in information theory. That is, suppose the network can process a block of k consecutive measurements and output k values of the function in, say, time $T(k)$. Then the computational throughput is $\frac{T(k)}{k}$. The block length k can be optimized, as can the policy for computing the k block of values for the function f. We call the reciprocal of the computational throughput the *computational cycle-time*.

We illustrate the results for the mean $f(x_1, \ldots, x_n) = \frac{x_1 + x_2 + \cdots + x_n}{n}$ and the max $f(x_1, \ldots, x_n) = \text{Max}_{1 \le i \le n} x_i$.

It runs out that the Max can be calculated in exponentially faster computational cycle-time. For example, the maximum computational throughput for the Mean in a random multi-hop network is $\Theta\left(\frac{1}{\log n}\right)$. On the other hand, the computational throughput for the Max is $\Theta\left(\frac{1}{\log \log n}\right)$.

The architecture for the computations exploits different ideas. For example, an order-optimal architecture for the mean is illustrated in Figure 7.19. The domain can be tessellated into small cells. In each cell, the total of the measurements can be calculated by a designated node. Then an in-tree rooted at the sink is used to pass on the total to the sink. There is really no advantage to block computation, at least vis-à-vis order optimality.

On the other hand, the computation of the Max can take advantage of block coding. To see this, let us consider a collocated scenario where all n nodes can all hear each other's transmissions. Suppose also for simplicity that there are only two values for the temperature, 0 and 1. Thus, any node with a temperature of 1 at a given time knows that it has the maximum temperature (possibly non-uniquely) among all nodes at that time.

Let us order the nodes as $1, 2, \ldots, n$. We now show how block computation can be exploited to increase computational throughput for the Max function. Node 1 collects a block of k measurements and announces only the list of times from among the k measurements at which its temperature is a 1. This list can be broadcast in an efficiently compressed manner. Node 2 then restricts attention to the set of times at which node 1 did not have a temperature of 1 and announces, within this subset, the set of times at which it had a temperature of 1. Again, this is also efficiently compressed. Node 3 then fills in furthers 1s, and the process continues. Due to the block coding, and due to the reduced time periods of interest as nodes progressively announce

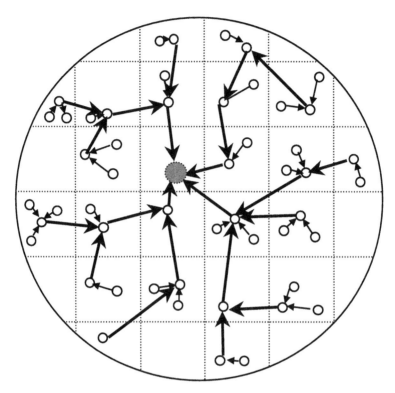

Figure 7.19 Order-optimal architecture for computing the Mean.

their times of temperature 1, the computational throughput is greatly improved.

Two classes, called "type-sensitive" and "type-threshold" functions, have been identified in Giridhar and Kumar[28] for which the computational throughput orders, as well as order-optimal architectures, have been determined.

7.6 Networked Control

As we noted in Section 7.3, the incorporation of *actuation* in addition to sensing gives rise to the ability to interact with the environment. This makes *control* over networks possible.

One of the key challenges, we believe, is to determine what are the appropriate abstractions and architectures. As an analogy, the network layer in the OSI stack (see Figure 7.6) manufactures the abstraction of a graph. This is then put to good use by the transport layer to manufacture the abstraction of a "pipe." This allows plug-and-play incorporation of protocols.

As another example of a proliferation-oriented architecture, one may consider the "von Neumann bridge" of serial computation.[29] Yet another is the source-channel coding separation theorem which lies at the heart of digital communication.[30] Then there is also the separation theorem of control[31] that divides the overall task into that of estimating the state of a system and then using that estimated state to calculate the control to be applied.

The question that arises is, see Figure 7.20, what are the appropriate abstractions for networked control? To investigate this we have built a testbed, shown in Figure 7.21. It consists of radio-controlled cars observed by video cameras. Each car is controlled by an individual laptop, and the laptops are connected by a wired/wireless ad hoc network.[32]

We have proposed one such abstraction that greatly facilitates networked control — the abstraction of *virtual collocation*.[32] To motivate this, we note that directly developing an application for a network where many sensors, actuators, and computational nodes are distributed is difficult. The measurement of time at different nodes is different since no two clocks agree.[33] Also, nodes such as cameras may need to be replaced, and one does not want to tie down the code

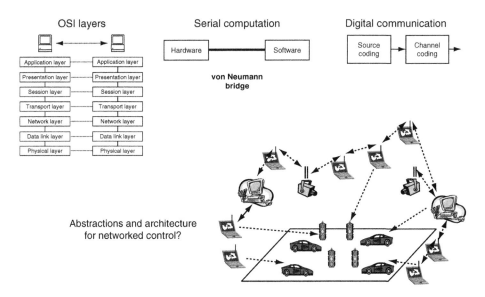

Figure 7.20 Challenge of architecture and abstractions for networked control.

Figure 7.21 The testbed in the Convergence Lab at the University of Illinois.

with issues related to addressing. One also wants to migrate function-ality from node to node automatically to optimize against latencies experienced while a system is running, i.e., at run-time. Essentially, one wants to simplify the development of applications for networked control.

For this purpose, we have developed a middleware for networked control, called Etherware.[34] It allows control designers to develop the application using components. It shields the control designer from having to know the computational resources on which the compo-nents are executing, their locations such as network addresses, or even the times as measured at differing components. Thus, it manufactures the abstraction of virtual collocation.

The net result is that the control designer can focus on the problem of control law development, rather than peripheral issues of imple-mentation. We believe that this is one of the keys to the future pro-liferation of networked control. We refer the reader to Reference 35 for videos of the system functioning in pursuit-evasion, traffic control, collision-avoidance, etc. modes.

The other key is a theoretical foundation for control algorithms that can provide stability, control loop performance, and robustness for a distributed system where not all information is known to all nodes and where what information is known to a node may be noisy or randomly delayed. This is a major challenge for the future.

References

1. Mobile Ad-hoc Networks (MANET), IETF Secretariat, 2005. http://www.ietf.org/html.charters/manet-charter.html.
2. N. Abramson, "The ALOHA system—another alternative for computer communications," *Proceedings of Fall Joint Computer Conference AFIPS Con.*, vol. 37, 1970, pp. 281–285.
3. D. Bertsekas and R. Gallager, *Data Networks*. Englewood Cliffs, NJ: Prentice-Hall, 1987.
4. C. E. Perkins and P. Bhagwat, "Highly dynamic destination-sequenced distance-vector routing (DSDV) for mobile computers," *Computer Communications Review*, vol. 24, no. 4, pp. 234–244, 1994. (ACM SIGCOMM '94.)
5. M. Horton, D. Culler, K. Pister, J. Hill, R. Szewczyk, and A. Woo, "MICA, the commercialization of microsensors motes," *Sensors*, vol. 19, pp. 40–48, April 2002.
6. P. Levis, S. Madden, D. Gay, J. Polastre, R. Szewczyk, K. Whitehouse, A. Woo, D. Gay, J. Hill, M. Welsh, E. Brewer, and D. Culler, "Tinyos: An operating system for sensor networks." To appear in *Ambient Intelligence*, Jan Rabaey, ed.
7. R. Szewczyk, J. Polastre, A. Mainwaring, J. Anderson, and D. Culler, "An analysis of a large scale habitat monitoring application," in *The Second ACM Conference on Embedded Networked Sensor Systems*, November 2004, Baltimore, MD.
8. K. Whitehouse, F. Jiang, A. Woo, C. Karlof, and D. Culler, "Sensor field localization: A deployment and empirical analysis," Technical Report UCB//CSD-04-1349, University of California, Berkeley, April 9, 2004.
9. D. A. Mindell, *Between Human and Machine: Feedback, Control and Computing before Cybernetics*. Baltimore: Johns Hopkins Press, 2002.
10. P. Gupta and P. R. Kumar, "The capacity of wireless networks," *IEEE Transactions on Information Theory*, vol. IT-46, pp. 388–404, March 2000.
11. A. Agarwal and P. R. Kumar, "Improved capacity bounds for wireless networks," *Wireless Communications and Mobile Computing*, vol. 4, pp. 251–261, 2004.
12. A. Agarwal and P. R. Kumar, "Capacity bound for ad-hoc and hybrid wireless networks," *ACM SIGCOMM Computer Communications Review*, vol. 34, pp. 71–81, July 2004. Special issue on Science of Networking Design.

13. S. Verdú, *Multiuser Detection*. Cambridge, UK: Cambridge University Press, 1998.
14. L. Xie and P. R. Kumar, "A network information theory for wireless communication: Scaling laws and optimal operation," *IEEE Transactions on Information Theory*, vol. 50, pp. 748–767, May 2004.
15. F. Xue, L. Xie, and P. Kumar, "The transport capacity of wireless networks over fading channels," *IEEE Transactions on Information Theory*, vol. 51, pp. 834–847, March 2005.
16. S. Narayanaswamy, V. Kawadia, R. S. Sreenivas, and P. R. Kumar, "Power control in ad-hoc networks: Theory, architecture, algorithm and implementation of the COMPOW protocol," in *European Wireless Conference – Next Generation Wireless Networks: Technologies, Protocols, Services and Applications*, (Florence, Italy), pp. 156–162, Feb. 25–28, 2002.
17. V. Kawadia and P. R. Kumar, "Power control and clustering in ad hoc networks," in *INFOCOM 2003* (San Francisco, CA), March 30–April 3, 2003.
18. V. Kawadia and P. R. Kumar, "Principles and protocols for power control in ad hoc networks," *IEEE Journal on Selected Areas in Communications*, vol. 23, pp. 76–88, January 2005.
19. C. E. Perkins, *Ad Hoc Networking*. Addison–Wesley, Reading, MA, 2001.
20. P. Gupta and P. R. Kumar, "A system and traffic dependent adaptive routing algorithm for ad hoc networks," in *Proceedings of the 36th IEEE Conference on Decision and Control* (San Diego, CA), pp. 2375–2380, December 1997.
21. J. G. Wardrop, "Some theoretical aspects of road traffic research," in *Proceedings of the Institute of Civil Engineering, Part 2*, pp. 325–378, 1952.
22. V. Borkar and P. R. Kumar, "Dynamic Cesaro-Wardrop equilibration in networks," *IEEE Transactions on Automatic Control*, vol. 48, pp. 382–396, March 2003.
23. V. Raghunathan and P. R. Kumar, "Issues in wardrop routing in wireless networks," in *Proceedings of WICON'05, First International Wireless Internet Conference* (Budapest, Hungary), pp. 34–41, July 10–14, 2005.
24. V. Raghunathan and P. R. Kumar, "Wardrop routing in wireless networks: From theory to implementation," in *Proceedings of the 43rd IEEE Conference on Decision and Control* (Bahamas), pp. 4661–4666, December 14–17, 2004. Invited paper.
25. R. Rozovsky and P. R. Kumar, "SEEDEX: A MAC protocol for ad hoc networks," in *Proceedings of the 2001 ACM International Symposium on Mobile Ad Hoc Networking and Computing* (Long Beach, CA), pp. 67–75, Oct. 4–5, 2001.
26. P. Gupta and P. R. Kumar, "Critical power for asymptotic connectivity in wireless networks," in *Stochastic Analysis, Control, Optimization and Applications: A Volume in Honor of W.H. Fleming* (W. M. McEneany, G. Yin, and Q. Zhang, eds.), pp. 547–566, Boston, MA: Birkhauser, March 1998.

27. P. Gupta and P. R. Kumar, "A system and traffic dependent adaptive routing algorithm for ad hoc networks," in *Proceedings of the 36th IEEE Conference on Decision and Control* (San Diego, CA), pp. 2375–2380, December 1997.

28. A. Giridhar and P. R. Kumar, "Computing and communicating functions over sensor networks," *IEEE Journal on Selected Areas in Communications*, vol. 23, pp. 755–764, April 2005.

29. L. G. Valiant, "A bridging model for parallel computation," *Communications of the ACM*, vol. 33, August 1990, pp. 103–111.

30. C. E. Shannon, "A mathematical theory of communication," *Bell System Technology Journal*, vol. 27, pp. 379–423, 1948.

31. W. M. Wonham, "On the separation theorem of stochastic control," *SIAM Journal on Control*, vol. 6, no. 2, pp. 312–326, 1968.

32. S. Graham, G. Baliga, and P. R. Kumar, "Issues in the convergence of control with communication and computing: Proliferation, architecture, design, services, and middleware," in *Proceedings of the 43rd IEEE Conference on Decision and Control* (Bahamas), pp. 1466–1471, December 14–17, 2004.

33. S. Graham, G. Baliga, and P. R. Kumar, "Time in general-purpose control systems: The control time protocol and an experimental evaluation," in *Proceedings of the 43rd IEEE Conference on Decision and Control* (Bahamas), pp. 4004–4009, December 14–17, 2004.

34. G. Baliga and P. R. Kumar, "Middleware architecture for federated control systems," Presented at Middleware 2003, Rio de Janeiro, Brazile, June 16–20, 2003. Published in IEEE Distributed Systems online, June 2003. http://dsonline.computer.org/0306/f/bal_print.htm, 2003.

35. IT Convergence Laborarory, "University of Illinois." http://decision.csl.uiuc.edu/~testbed/.2005.

Index

Printed and bound by CPI Group (UK) Ltd, Croydon, CR0 4YY

03/10/2024

01040412-0008